To dear Joyce
With Best Wishes.
love
George.

Thank you!

DID I TAKE THE RIGHT TURNING?

Group Captain George F.K. Donaldson
DFC AFC FRMetS FRAeS

Did I Take
The Right Turning?

by

George Donaldson

The Pentland Press Limited
Edinburgh · Cambridge · Durham · USA

© George Donaldson 1999

First published in 1999 by
The Pentland Press Ltd.
1 Hutton Close
South Church
Bishop Auckland
Durham

British Library Cataloguing in Publication Data.
A catalogue record for this book is available
from the British Library.

ISBN 1 85821 672 9

Typeset by George Wishart & Associates, Whitley Bay.
Printed and bound by Bookcraft Ltd., Bath.

Contents

Illustrations

Preface

When one or two friends suggested that I write my memoirs I pooh-poohed the idea. They pointed out that history is largely made up of the lives and experiences of people and that everyone had some part that was unique. They said that I had lived through quite momentous times, including two world wars, the events which led up to them, and those which succeeded them. I have therefore put down a few bits as factually as possible. They may at least be of some interest to members of 'The Clan', who may come after me!

Speaking of 'The Clan', these memoirs would never have got off the ground without the help and encouragement of my cousin Jim Gilmour who typed and retyped them several times; and urged me to finish them off, when I was losing interest.

CHAPTER 1

Early Days

I was born on 24 February 1908, in Wallasey, Cheshire (Merseyside). My father and mother were Scots, Father being born in Leith, Edinburgh, Mother in Greenock. Father was an engineer by profession; he and Mother came down to England when Father got a job with the Mersey Docks and Harbour Board. We moved to Birkenhead circa 1910. I had two older sisters, Agnes and Jane, seven and four years older than me respectively. We were a normal happy family, staunch members of the local Presbyterian church, Father who had a fine tenor voice, and Mother, a contralto, singing in the choir. I sang too – until my voice broke!

I went to school at the Birkenhead Institute, a secondary school. In those days one could enter through passing a scholarship from an elementary school – or by paying fees. I sat the scholarship, but did not pass; so Father had to pay the fees, which I recall were three pounds fifteen shillings per term. In retrospect I suppose it was a good school, though alas, it no longer exists. At the age of fifteen I sat the examination of the Northern Universities, known as School Certificate. If one obtained a 'Credit' in six subjects, one matriculated. Of these, Maths and Latin were compulsory. I only achieved a 'Pass' in them, which barred my going on to university – which Mother and Father would have liked.

The school played soccer and I was privileged to be in the team which won the Liverpool and District Shield in 1921/22. I recall that we played the final on Everton's ground. I also played cricket for the school, but was never very good.

I left school at the age of sixteen without any particular ambitions. Mother and Father would have liked me to become a doctor; they also thought I might make a lawyer – even in those days I liked an argument! But lack of Matriculation ruled them out. A number of my friends were in banks or insurance companies. I applied for several, but alas, Matriculation was needed. Then our church came to the rescue. Our 'Elder', a Mr Proctor, was a director of a number of companies in Liverpool. The poor chap went blind. Mother, ever an opportunist, asked him if he would like me to be his guide. 'I'd be most grateful,' he said. So my first job was 'Leader of the Blind'.

The job took us around a number of companies: cotton, then in its heyday in Liverpool, provisions, fruit, and insurance. One day we attended a meeting at the State Assurance Company in Dale Street. Descending the rather grandiose staircase he paused to survey the General Office, with its serried ranks of clerks perched on high stools at long desks, and said, 'Donaldson, would you like a job here?' 'Yes Sir' I replied, and became one of them. How true is the saying, 'It's not what you know, it's who you know!'

I was initially in the Fire Department. I found the work quite interesting, and studied for the Associateship of the Chartered Insurance Institute. Surprisingly, I managed to pass Part One – with Honours. Then I was moved – to the Cash Department. It was supposed to be a 'step up'. There were four in the department – the Chief Cashier, the Assistant Cashier, the Senior Cashier, and myself. Apart from helping with the books, and taking my turn at the counter, I had to go out and call on firms which were slow in paying their accounts. A fair number of these were cotton firms, in which my friends worked. Quite often, if it was near lunchtime, they would suggest 'going for a pint'. I didn't need much persuasion, and was duly 'ticked off' when I got back. My excuse was, 'Well, I got the money!'

On the leisure side I was keen on sport, and developed a liking for rugby in preference to soccer. I joined a junior club, Bidston, and got into the first team. I suppose I was what is called 'spotted' by one, Jock Montador, a member of Birkenhead Park and a Scottish trialist. He invited me to join 'Park'. I didn't need much persuasion. In summer I played cricket and tennis, but wasn't very good at either. I joined the Oxton Cricket Club largely because most of my rugby friends were members. It was more of a 'jolly' than serious cricket, and formed a nice prelude to the evening's festivities.

I recall the General Strike – in 1925. All transport came to a standstill, so that we could not get over to Liverpool. We did think of getting a boat and rowing across the Mersey, 'just for the heck of it', but it was pointed out that the currents were very treacherous and we abandoned the idea. Then we heard there were ships in the Birkenhead docks with vital food on board, so we volunteered to unload them. My job was to sit on the dockside at the end of a chute from a grain ship. Sacks of grain would come down the chute; with a sharp knife, I would slit the mouth of the sack, and upend it through a manhole onto a travelling band which delivered it to the granary. At the end of the day we were faced with the problem of getting out of the dockyard. The main gates were picketted by angry strikers; we had to find a quiet section of the dock-wall and shin over it. Even so, we had several skirmishes. I recall one occasion when we were larking about and I got shoved into the dock. It was about ten feet to the water – which was filthy; then I had to swim about fifty yards to the nearest steps.

The General Strike ended – alas, without the strikers achieving what they had gone on strike for. So it was back to my desk at the office. My salary was £60 per annum.

CHAPTER II

Airborne

The year was 1929. I was twenty-one years old, and working as a clerk in the State Assurance Company, in Liverpool. I was initially in the Fire Insurance Department, but had recently been transferred to the Cash Department. After four years my salary was £90 per annum.

The phone rang. Could I turn out for a scratch Birkenhead Park rugger team to play against RAF Sealand on Wednesday? Life was fairly relaxed in the firm. I asked my boss – and he said 'Yes'.

The game – and its result – were immaterial: what interested me was the entirely new life which the visit revealed. RAF Sealand was the base for No. 5 Flying Training School, which trained Short Service Commissioned officers, who then formed a large proportion of RAF pilot requirements. The term of service was five years, followed by four years on the Reserve. The starting salary was £240, rising to £450. There was also a gratuity of £250 on completion of the five-year period.

After the game we were invited to tea in the Officers' Mess, followed after a brief interval by an invitation to 'have a pint, old boy'. Most of the team were SSC officers in their first year. In conversation I learnt a bit more about their life.

When I got home I thought about my visit and the glimpse of life in the RAF which it had revealed. I discussed it with my friend Hubert who worked with me. Together we decided to get particulars and apply for Short Service Commissions. Looking back honestly, it was not the excitement of flying, but the glamour of the life – and the greater salary – which attracted me.

The application forms were searching as to one's schooling, academic qualifications and sporting activities. In due course Hubert and I received invitations to report to Air Ministry HQ, London for interview. We had jumped the first hurdle.

The visit to London was an event in itself. I had never been there before. I recall being told by some well-wisher, 'If they ask you why you want to join, don't say, "It may enable me to get a job in Civil Aviation afterwards"! They want people dedicated to Service flying.'

After the interview came the Medical Board. I had not been worried about this, but after blood and urine tests was told that sugar had been found in my urine, and that I would have to have an additional test, called a 'sugar tolerance test'. This involved taking a sugar solution at intervals and the doctors taking samples of my blood and urine. By plotting these they were able to ascertain that I had a condition known as 'Renal Glycosuria' which allowed some of the sugar occasionally to 'spill over'. Recent progress in medical research had shown that this was not harmful. A few years earlier I would have been turned down as a diabetes suspect!

I was accepted for a Short Service Commission. For some reason which I shall never understand, my friend Hubert was not. I was very sorry.

Uxbridge

My joining instructions were to report to RAF Depot, Uxbridge on 11 April 1930. I went by train from my home in Birkenhead, to London Euston, then by underground to Uxbridge, and on foot to the Depot.

Our intake comprised twenty-six chaps between the ages of eighteen and twenty-six, from varied walks of life. One of the recruits was thirty-one; an exception was made in his case because he had served as a pilot in the Royal Flying Corps.

We were at Uxbridge for two weeks, during which we received a general introduction to Service life, its organization, basic drill, relative ranks – who had to be saluted (in mufti one raised one's hat) – and were fitted out with uniform. We started our drill in the clothes we came in. We must have looked a motley crew, variously attired in grey flannels, plus-fours, pinstripes, city suits, bowler hats, trilbys, caps...!

Our scale of uniform was elaborate:

Working Dress	– Barathea tunic with belt.
	Slacks, white shirt with soft collar, black tie, socks and shoes.
	Greatcoat with belt.
	Peaked cap.
Ceremonial Dress	– Breeches and puttees, boots, stiff white collar.
Mess Kit	– Short monkey jacket with epaulettes.
	Overalls (tight trousers).
	Long patent-leather boots.
	Black bow tie, stiff-fronted white shirt.
Miscellaneous	– Walking stick.
	Brown leather gloves.

We were supplied with a list of approved tailors with their brochures and prices. These included such well-known firms as Moss Bros., Alkit, Poland, Gieves etc. I chose Gieves. The cash allowance was £50. If one could get the kit for less – by shopping around – one was entitled to keep the balance. I recall that I saved three or four pounds.

Thinking back I wondered if I had made a mistake in the prices – or that they would be believed today. So I wrote to Gieves, and received a most courteous reply from the Vice-Chairman, Robert Gieves, confirming my recollections. A further letter estimates the present-day costs of the kit at £1,907! (See illustrations of correspondence and brochure.) Having got one's uniform from a bespoke tailor, it was normal to open an account and buy other clothes. I had never had a tailor before, and soon ran up an account which proved embarrassing. Gieves also sold fine jewellery. Later when I met my dear Sue she was to have gold 'Wing' brooches and other RAF crested items.

Towards the end of our fortnight at Uxbridge we were informed that we were to be posted to No. 3 Flying Training School, Grantham (later to revert to its First World War name, Spittalgate).

Grantham

The course which was to train us up to 'wings' standard was divided into two terms, Junior and Senior, of six months each. The Junior term was devoted to '*ab initio* training'. Our aircraft was the Hawker 'Tom Tit'. This delightful aircraft was one of three selected to replace the legendary Avro 504 which had been in operation since the First World War. The other two were the Avro 'Tutor' and de Haviland 'Gipsy Moth' DH 80 – destined to become even more legendary under the label 'Tiger Moth'. The 'Tom Tit' lost out, but the other two were both brought into operation. Nevertheless it is very obvious, in retrospect, that the 'Tom Tit' was the forerunner of the famous line of Hawker aircraft: 'Hart', 'Fury', 'Hind', 'Demon' etc.

In my Junior term, under the expert and patient tutelage of P/O A.E. Dark, an ex-Halton apprentice, I went solo in fifteen hours. My total flying time in the Junior term was forty-seven hours, of which twenty-five were solo.

GIEVES & HAWKES
No. 1 Savile Row, London

Please reply: No.1 Savile Row, London W1X 2JR Telephone: (071) 434 2001 FAX: (071) 437 1092

RGWG/JG/Misc

9th October 1991

Group Captain G.F.K.Donaldson.R.A.F.
18 Park Road,
West Kirby,
Wirral,
L48 4DW.

Dear Group Captain Donaldson,

I have been intrigued to read your letter of 17th September with recollections of times past when Gieves provided your initial R.A.F outfit.

It so happens our archive contains a 1931 price list, and this I have photocopied and now enclose. Reference to page 15 will show that, after discount, there were indeed a few pounds change from £50.00.

I wish you well with the memoirs.

Yours sincereley,

Robert Gieve
Vice Chairman

TWO CENTURIES OF FINE TAILORING

Gieves & Hawkes Limited, Registered in England No. 102640 Registered Office: 1 Savile Row, London W1X 2JR
Gieves & Hawkes International Limited, Registered in England No. 1800278, Registered Office: 1 Savile Row, London W1X 2JR

Gieves & Hawkes letter – 9 October 1991.

GIEVES & HAWKES
No.1 Savile Row, London

Please reply: No.1 Savile Row, London W1X 2JR Telephone: (071) 434 2001 FAX: (071) 437 1092

14th October 1991

Group Captain G F K Donaldson RAF
18 Park Road
West Kirby
Wirral
L48 4DW

Dear Group Captain Donaldson

I can only partially answer your question of 11th October, because the circumstances of outfitting are now very different. A student officer now receives all his uniform outfit from a contractor at no expense to himself, save as a tax payer. However, the Gieves & Hawkes full bespoke prices for the main items are now as follows:-

RAF Best Barathea Jacket and Trousers	£870.00
P/O Braid	£ 11.00
RAF Mess Jacket and Trousers	£915.00
P/O Lace	22.00
Waistcoat Optional	-
Cummerbund through stores	
RAF Cap and Badge	£ 89.00
	1907.00

Today, most of our RAF business is for the middle to senior echelons of the Service, who require a decent uniform for a particular appointment.

Yours sincerely

Robert Gieve
VICE CHAIRMAN

TWO CENTURIES OF FINE TAILORING

Gieves & Hawkes Limited. Registered in England No. 1026430 Registered Office: 1 Savile Row, London W1X 2JR
Gieves & Hawkes International Limited. Registered in England No. 1900278. Registered Office: 1 Savile Row, London W1X 2JR

Gieves & Hawkes letter – 14 October 1991.

9

GIEVES, LTD.

ROYAL AIR FORCE

OUTFIT LIST FOR SHORT SERVICE COMMISSION

1. SERVICE DRESS

			£	s	d
REGULATION BOOTS	.	1 pair	1	10	
" SHOES	.	1 pair	1	10	
UNIFORM CAP AND BADGE	.		1	7	
TAN REGULATION GLOVES	.	1 pair	0	7	
BLUE NAP REGULATION GREATCOAT	.		9	10	
PUTTEES	.	1 pair nett	0	12	
REGULATION TUNIC	.		6	17	
" BREECHES	.	1 pair	4	10	
" SLACKS	.	1 pair	2	15	
BLACK UNIFORM TIE	.		0	3	
REGULATION CANE	.		0	7	
GYM. SHOES	.	1 pair	0	4	

2. MESS DRESS

			£	s	d
MESS CAP AND BADGE	.		1	12	
SUPERFINE MESS JACKET (Complete)	.		8	10	
" WAISTCOAT	.		2	2	
MESS OVERALLS	.	1 pair	3	7	
PATENT LEATHER FULL WELLINGTON BOOTS	1 pair		2	10	
WHITE DRESS GLOVES	.	1 pair	0	5	
BLACK DRESS TIE	.		0	3	
MESS STRAPS FOR GREATCOAT	.	1 pair	0	12	
			£48	18	
		Less 5 per cent. discount	2	8	
			£46	9	

OTHER ESSENTIAL ITEMS

			£	s	d
WHITE DAY SHIRTS	.		from £0	5	
" DRESS SHIRTS	.		" 0	10	
STIFF DOUBLE COLLARS	.	from per doz.	0	10	
SOFT COLLARS	.	"	0	10	
DRESS COLLARS	.	"	0	2	
BLACK CASHMERE SOCKS	.	from per pair	0	5	
" DRESS SOCKS	.	"	0	2	
REGULATION WATERPROOF COAT	.		2	10	
TIN CASE	.		3	7	
PAINTING NAME ON SAME	.	Nett	0	2	

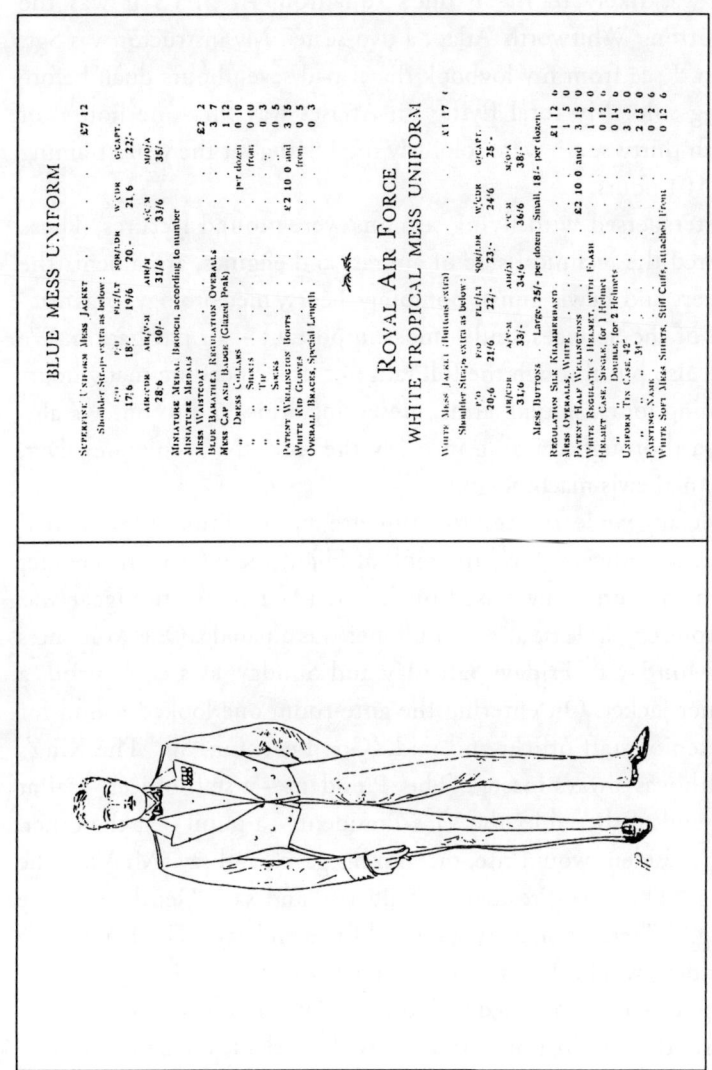

BLUE MESS UNIFORM

Superfine Undress Mess Jacket . . . £7 12

Shoulder Straps extra as below:

F/O	F/L	PLT/LT	SQDN/DR	W/CDR	G/CAPT.
17/6	18/-	19/6	20/-	21/6	22/-
AIR/CDR	AIR/V-M	AIR/M	A/C-M	MD/A	
28/6	30/-	31/6	33/6	35/-	

Miniature Medal Brooch, according to number
Miniature Medals . . . £2 2
Mess Waistcoat . . . 3 7
Blue Barathea Regulation Overalls . . . 1 12
Mess Cap and Badge (Glazed Peak) . . . 1 12
Dress Collars . . . per dozen 0 10
   " Shirt . . . from 0 3
   " Tie . . . 0 5
   " Socks . . . 0 3
Patent Wellington Boots . . . £2 10 0 and 3 15
White Kid Gloves . . . from 0 5
Overall Braces, Special Length . . . 0 3

ROYAL AIR FORCE
WHITE TROPICAL MESS UNIFORM

White Mess Jacket (Buttons extra) . . . £1 17

Shoulder Straps extra as below:

F/O	F/L	PLT/LT	SQDN/DR	W/CDR	G/CAPT.
20/6	21/6	22/-	23/-	24/6	25/-
AIR/CDR	AIR/V-M	AIR/M	A/C-M	MD/A	
31/6	33/-	34/6	36/6	38/-	

Mess Buttons . . . Large, 25/- per dozen: Small, 18/- per dozen.

Regulation Silk Kummerbund . . . £1 12 6
Mess Overalls, White . . . 1 5 0
Patent Half Wellingtons . . . £2 10 0 and 3 15 0
White Regulation Helmet, with Flash . . . 1 6 0
Helmet Case, Single, for 1 Helmet . . . 0 12 6
   " Double, for 2 Helmets . . . 0 14 0
Uniform Tin Case, 42" . . . 3 5 0
   " 31" . . . 2 15 0
Painting Name . . . 0 2 6
White Soft Mess Shirts, Stiff Cuffs, attached Front . . . 0 12 6

RAF outfit list for Short Service Commission Officer.

11

The Senior term was devoted to training on the larger aircraft one was likely to fly in one's squadron. At 3FTS it was the Armstrong Whitworth 'Atlas', a two-seater. My instructor was Sgt. Ginn. I see from my logbook that I had seven hours dual, before flying solo. My total flying on Atlases was fifty-one hours, of which thirty-seven were solo. My total flying for the year's training was 101 hours.

Interspersed with flying sessions were ground lectures. These covered the technical side of aircraft and engines, armament (the Vickers and Lewis guns), bombing theory, meteorology, organization of the Services, and – most important – air navigation. We were also put through the full range of drill, spending many hours forming fours, sloping arms, presenting arms, and so on. We also had a thorough training in firing the rifle, the Webley revolver, and the Lewis machine-gun.

Socially we learnt Service etiquette and tradition. One saluted all senior officers above the rank of Flight Lieutenant. If one met them in mufti, one raised one's hat. (As I recall, headgear was compulsory, at least at FTS.) Dinner was a parade. One wore mess kit Monday to Friday: Saturday and Sunday, as a concession, a dinner jacket. On entering the ante-room one looked round for the senior staff officer and said, 'Good evening, Sir.' The King's health was always toasted. The President – a staff officer – sat at one end of the table, the Vice-President – a pupil – at the other. The President would rise, pick up his gavel, and say, 'Mr Vice, the King.' The Vice-President would rise and say, 'Gentlemen, the King.' Whereupon everyone would rise and say, 'The King.' The President would then say, 'Gentlemen, you may smoke.'

Sport was encouraged –Wednesday afternoon was compulsory sport. If one did not play the usual football, cricket, tennis or squash then one had to go for a cross-country run. I played rugby and cricket for the station, though I was not much of a cricketer. I

12

also played squash for the first time and became addicted to it. One of our rugby fixtures was against Nottingham 'A' team. We beat them rather handsomely. Following the match I received an invitation to play for Nottingham. This of course meant going over there on Saturdays, and travelling around with the team. I got permission, and much enjoyed my short spell playing for Nottingham. Then one day I got a card from the RAF RU to play for the representative team against the United Banks. Oddly I played not full back, my normal position, but centre.

One of our outstanding players was Kenneth Cross. He was a very fine wing-threequarter. I have always been puzzled as to why, if I was selected to play at centre, he was not automatically selected as my wing! However this was soon rectified – 'Bing', as he was christened, thereafter played regularly for the representative RAF team, eventually becoming Captain.

'Bing' went on to achieve distinction. After commanding No. 44 Squadron in the ill-fated Norwegian campaign, he saw service in Europe, the Middle East and the Far East – followed inevitably by Staff posts in the higher echelons of the Ministry. After retirement – as Air Chief Marshal Sir Kenneth Cross, KCB, CBE, DSO, DFC, and many allied decorations – he worked for the Red Cross. I count myself privileged to have served with him.

One of the most important subjects in our training was navigation, the ability to find one's way from A to B. After class training in theory, meteorology, maps and charts, laying off courses etc. one had to complete four cross-country flights. Each of mine had its own story, but I will mention only the final one, Grantham to Sealand, in any detail. The weather was poor, and I cannot say that I always knew where I was, but the finding of Sealand was easy owing to its situation near the River Dee and the Wirral Peninsula – which was my home territory. The return flight was a different matter. I soon lost my place on the map, but held

'As we are'.

Hawker 'Tom Tit' two-seater biplane.

my course, hoping to identify familiar country around Grantham. Unfortunately I had forgotten about the onset of darkness. I could not locate myself, so decided I would have to find a suitable place to land. I found a field which I thought I might, with luck, get into. I made a couple of passes at it, and then said 'Here goes'. I touched down, and soon saw the hedge looming close. The 'Atlas' had no brakes: the only way one could reduce speed and shorten a landing run was by kicking the rudder violently from side to side. I came to rest a few yards from a deep ditch alongside the hedge. I turned off fuel and ignition, as instructed, and looked around. I spotted a house nearby, and started to walk towards it, to be met by a boy running towards me excitedly. It was his house, and he invited me to go back with him and use his phone. I found that I was near Bourne, about fifteen miles south-east of Grantham. I got through to the orderly officer and told him what had happened and where I was. 'You're the fourth,' he said. He told me to make my way into Bourne and wait by the market cross for a relief party.

After an hour or so a van arrived with a flight sergeant and three airmen.

'Right, take us to the plane,' he said.

It was quite dark by this time, and I could not find the field. Eventually I spotted the house, and was able then to retrace the route. An airman was left on guard, and I was taken back – to face interrogation. I learned to my slight consolation that some of my colleagues had not been so fortunate, and had written off their aircraft in trying to land. One of them, a colourful Irishman, landed near a large house and was given hospitality by the owner, who had an attractive daughter. 'JGB', as he was known, made quite a story of it!

At the end of the year's course, during which I had flown 101 hours, came the CFI's test. It was a somewhat daunting experience, but I passed. The categories available – and to be stamped in one's logbook – were 'Below Average', 'Average', 'Above Average', and 'Exceptional'. I was categorized 'Average'.

Towards the end of our time, a list was displayed, inviting us to state which type of operational squadron we would each prefer to join. The choice was 'Fighter', 'Bomber', 'Army Co-operation', or 'Coastal'. The glamour, of course, lay with the fighters. I put my name down for bombers. I was posted to Army Co-operation with No. 2 (AC) Squadron. I received a few 'Bad luck, old boy's – but I was never to regret it.

CHAPTER III

Army Co-operation

Manston was the home of No. 2 (Army Co-operation) Squadron. Our aircraft was the 'Atlas' which I had flown on my advanced course. I did a little local flying, but had to await an Army Co-operation Course before being able to participate usefully in the squadron's role.

Situated in the Isle of Thanet, Manston was well served with holiday resorts. Margate, Ramsgate, Birchington, Herne Bay were all within a few miles. At Birchington was a hotel called the Beresford, popular with weekend visitors, some of whom used to arrive in private aircraft at Manston. On Saturday night the hotel ran the inevitable dinner-dance. The officers of Manston had a standing invitation to attend after dinner. The idea was that we provided partners for unattached daughters who came with their parents. I recall one occasion being met by the hostess, who took me by the arm and said, 'Pilot Officer Donaldson, I would like to introduce you to a rather special family of guests with us tonight.' She took me over to a table where a man and his wife were sitting with their daughter, whom I judged to be about eighteen. 'Mr and Mrs Byass, I'd like you to meet Pilot Officer Donaldson.' I was invited to join, introduced to their daughter, Pam (I think)) and handed a glass of wine. After a short interval, the band struck up, and I asked Mrs Byass for a dance. In these circumstances one always danced with mum first! We had a pleasant evening, during which I suggested that they might like to have a look over the aerodrome the next morning, and have lunchtime drinks in the mess. They said they would be delighted.

The next morning I was casually watching out when a gleaming chauffeur-driven Rolls Royce came into the drive. Shortly afterwards a mess waiter approached and said, 'Your guests have arrived, Sir.' You are probably ahead of me. Yes – he was indeed the senior partner in Gonzales Byass, the well-known sherry importers. They gave me a cordial invitation to visit them – I never did. Did I miss a turning?

Ramsgate was the home of the prestigious Royal Temple Yacht Club, of which the Officers' Mess was privileged to be corporate members. Indeed we had our own boat – a dinghy of the 'Whitstable' class. I had never done any sailing; it was a great pleasure to learn.

One Sunday three of us were manoeuvring our boat out of the harbour. In those days there used to be a service by paddle-boats from London to Ramsgate. As we were manoeuvring our boat out of the harbour – a tricky business at the best of times – the paddle-boat suddenly started up: obviously the Captain had not seen us. We found our boat being drawn towards the thrashing paddles. At the last minute we managed to contact the leading edge of the paddle box and push the boat clear. It was an experience I shall always remember.

The role of the Army Co-operation Squadron was varied and interesting. Much of the success of an army depends on its ability to know the disposition, strength and composition of the forces opposing it. Traditionally this was achieved by ground reconnaissance, observation from hills – and of course by spies. Later, balloons came into being, enabling the Army to see a little further, but only from the confines of their own lines. The advent of the aeroplane enabled observers to penetrate enemy territory and report – at the risk, of course, of being shot down by enemy aircraft or gunfire. Of particular value was the ability of an observer in an aircraft to report the position of artillery shots relative to the target, and thus allow corrections to be made. Radio telephony was in its infancy in those days. The standard method of communicating with the Army was by a combination of Morse Wireless Telegraphy, ground signals and the picking up and dropping of messages.

Photography was another important role. It covered vertical shots by a camera mounted in the aircraft, which produced mosaics – a series of vertical shots which overlapped to produce a map of an area, and oblique shots taken by the observer with a hand-held camera.

To afford some means of protection against enemy fighters and ground defences, the aircraft was equipped with a Lewis gun mounted on a scarf ring, operated by the observer/gunner/WT operator. There was also a Vickers gun, fixed to fire ahead through the orbit of the propeller. It was prevented from hitting the propeller blades by a most ingenious device known as Constantinesco gearing. The aeroplane was also fitted with bomb racks under each wing. Bombs could be released by the pilot in a diving attack, or from higher altitudes by the observer, using a precision bomb-sight and giving his directions to the pilot.

Thus the Army Co-operation pilot carried out the role of fighters and bombers, in addition to its main task of helping the ground forces.

All these duties had to be learned – one was of little use to one's squadron until they had been – and in due course I was posted to the School of Army Co-operation, Old Sarum, Wiltshire.

Old Sarum

Old Sarum is very close to the historic mound of that name, which was the old city of Salisbury, now some three miles to the south. I renewed acquaintance with a number of old colleagues, which made my time here very pleasant. I will not go into detail on the training for the roles I have mentioned, but the training in aerial photography and in artillery shoots are perhaps worth a mention.

In photography our first vertical pinpoint was, universally, of Stonehenge, about ten miles away. One was guided over it by one's observer, who was lying prone on the floor peering through a hole and giving instructions – 'Right', 'Left, left' (left always repeated to avoid mistakes), 'Steady', 'Steady', 'OK'.

Vertical mosaics needed a lot of pre-flight preparation, in which the Met wind forecast played an important part. Firstly the area was marked off on the map, and runs – tracks to be flown – were drawn, spaced such that the correct lateral overlap (30 per cent) was achieved. If possible tracks were made on line with the forecast wind, so that there was little drift. If this was not possible, then a course had to be worked out, which compensated for drift. Any drift not allowed for would stagger or 'step' the line of photos. Similarly ground speed had to be worked out so that photos were taken at the right interval to obtain the desired vertical overlap. After a photographic reconnaissance one could not get to the photographic section quickly enough to see the results – which were displayed for all to see, and thus generated great rivalry.

Co-operation with the Artillery involved a prescribed procedure. The battery carrying out the 'shoot' would display a ground signal when ready to start. The pilot would then position his aeroplane over the target, and when he was in a good position to observe it, would tap out on his morse key the order to fire 'Kay tock, kay tock' (KT). The shell would take perhaps 10-15 or more seconds to reach the target, depending on the range. This would have been worked out before the shoot, and the pilot had to count it accurately so that he was in the optimum position to observe the 'burst'. Observations were related to an imaginary clock which the pilot superimposed on the target, with twelve o'clock always to the north. Distance from the target was according to a code – A 50 yards, B 100 yards, C 200 yards and so on. Thus, if the pilot estimated that the burst had fallen 200 yards to the east of the target, he would tap out to the Battery Commander 'C3, C3'. On receipt of this, the Battery Commander would not immediately apply a correction to bring his fire 200 yards to the west, because the first shot might have been a random one. He would 'lay off' 400 yards to the west. If the pilot then reported 'C9, C9' the Battery Commander would have 'bracketed' the target, and would then lay off 200 yards to the east, which theoretically – and quite often in practice – should hit the target. Usually it was not quite so cut and dried, and several further corrections were given. When a burst was eventually on the target the pilot would signal 'OM, OK, GO, GO'. The Battery would fire as quickly as they could on their last siting and range.

The school had devised a most useful and ingenious synthetic trainer for practising this procedure without leaving the ground. It was based on the use of a squash court! The pilot sat up in the gallery with his morse key and map. The whole area of the squash court below was covered with a large-scale map at a height of about six feet from the floor, such that one could walk underneath the map. Several targets were pre-determined, and holes were made around them at designated positions and distances. At the start of the 'shoot' the 'gunner' would position himself underneath the map, with a lighted cigarette. He would identify the range, and wait for the order 'Fire' (KT, KT) from the pilot. He would then estimate the time taken for the shell to travel, and then, having drawn on his cigarette, puff a mouthful of smoke through one of the holes. To the pilot in the gallery it was most realistic. He would report its position relative to the 'clock' code, as set out above, and the 'Gunner' poised with his cigarette beneath the map would simulate the correction with the appropriate puff. The trainee pilot in the gallery with his morse key would report the burst as if he were in his aeroplane – without risk.

Another manoeuvre which we had to learn – and which, indeed, required a lot of practice before one became confident, was Message Picking Up – or MPU as it was called. This was a means by which the Army on the ground could pass a message to a reconnaissance pilot. Two rifles with bayonets attached were stuck into the ground – by the bayonets, of course. They were positioned about ten feet apart, and the message in a bag was strung between the butts of the rifles – and so was about four feet above the ground. Underneath the aeroplane was a rod about eight feet long with a hook at the end, which was let down by the observer. When working with an Army unit, one knew the location of the particular HQ, and flew over it periodically to see if any signals were displayed. If the signal denoting 'We have a message for you' was displayed, one flew round, the observer lowered the hook, and one flew towards the outstretched line containing the message, gauging one's height so that the hook engaged the line and snatched off the message. The A/C/Observer then pulled in the hook, disengaged the message, and passed it forward for the pilot to read. It was quite a tricky operation, and rifles were sometimes damaged. It was not always successful first time, either. One of our training exercises was to make six attempts at the manoeuvre. For these we had wooden uprights in place of rifles. If one got 'six out of six' one was doing very well.

I returned to Manston, a trained but inexperienced Army Co-operation pilot.

Our Station Commander was Group Captain Pink. One day I was told to report to his office. I wondered what I had done amiss! 'Donaldson,' he said, 'I have long felt that our present uniform is

GIEVES & HAWKES
No.1 Savile Row, London

Please reply: No.1 Savile Row, London W1X 2JR Telephone: (071) 434 2001 FAX: (071) 437 1092

Group Captain G.F.K. Donaldson R.A.F.,
18 Park Road,
West Kirby,
Wirral,
Lancs.
L48.4DW.

Dear Group Captain Donaldson,

Your recollections of the R.A.F. Uniform Trials at R.A.F. Manston in 1931 make most interesting reading. I had no idea that such changes had been contemplated. However it comes as no surprise that Gieves were envolved in the venture for we had played a full part in the evolution of R.A.F. Uniform from the days of the R.F.C.

Thank you for the loan of the photographs which I now return. I have taken copies for the archives.

Yours sincerely

R.J.W.GIEVE
Vice Chairman

TWO CENTURIES OF FINE TAILORING

Gieves & Hawkes Limited. Registered in England No. 102640. Registered Office: 1 Savile Row, London W1X 2JR
Gieves & Hawkes International Limited. Registered in England No. 1900278. Registered Office: 1 Savile Row, London W1X 2JR

RAF officer's uniforms in the early 1930s.

24

The 'new' uniform.

impracticable and could be improved. I have some modifications to propose, and I would like you to 'model' them. Gieves' tailor will be coming down to measure you up.' His proposed modifications were:

1. To abandon the breeches and puttees on the grounds that they were cramping to one's movements when getting in and out of an aircraft.
2. To substitute for them 'plus-fours' and stockings.
3. To replace the current white shirt by a blue one.
4. To replace the peaked cap – difficult to stuff away when flying – by a forage cap (as in the former Royal Flying Corps) and a beret.
5. To introduce a 'Sam Browne' leather belt and a silver knobbed cane for ceremonial occasions.
6. To abolish the stiff white collar except for mess kit (when it would be 'butterfly').

I was duly measured up, together with an airman whose uniform had been similarly modified. We were paraded at the next AOC's inspection. Items 1, 3, 4, & 6 were later adopted.

After a month or so back at No. 2 Squadron, Manston, I found myself posted to No. 20 (Army Co-operation) Squadron, Peshawar, North-West Frontier Province, India. At the time my feelings were mixed, and I received a certain amount of commiseration. But I have never regretted it.

CHAPTER IV

India

In those days trooping was by sea. After all the formalities were completed – inoculations, tropical kit etc. – I reported to Southampton to join the P & O troopship *Neuralia*. The RAF contingent comprised a number of old colleagues, two destined for No. 20 Squadron. But most of the troops were Army. Among them were some with 'ULIA' on their shoulders. This, I found, indicated 'Unattached List Indian Army'. Officers destined for the Indian Army served a year with a British Army regiment before being accepted by an Indian Army regiment. The ship sailed via Gibraltar, across the Mediterranean, Suez Canal, Red Sea, Aden across the Arabian Sea, to Karachi. At Karachi we boarded the famous 'Frontier Mail'. It had about four classes, not including the tops of the carriages, which were the normal method of travel for many 'locals'. We travelled first class. The carriages were virtually suites, and very comfortable. The journey took a day and a half through country rather drab, and made more so by the number of beggars constantly holding out their hands at every stop. I was met at Peshawar by an old friend, Philip Haynes.

Peshawar is an historic city lying athwart the famous Khyber Pass route through Afghanistan to Russia. Most of the Army, Diplomatic and European businessmen lived in the cantonment outside the city. It spread either side of the Mall, a wide, tree-lined avenue on which most of the officers' messes were situated, fronted by attractive lawns and gardens. Units of the British and

Indian Armies rotated every few years. There was always a British and an Indian cavalry regiment and infantry regiments. I recall the Gordon Highlanders, King's Own Yorkshire Light Infantry, 18th Cavalry, Sikhs, Punjabis, Poona Horse, as well as supporting services, Signals, Royal Engineers, Medical Corps etc. The Army was of Brigade strength. While I was there the Brigade Commander was Brigadier Claude Auchinleck, later to achieve the highest rank and serve with distinction in the Second World War.

The aerodrome was just outside the cantonment; it was naturally of the hard-baked earth indigenous to the country, but grass had been sown and it was constantly watered, and so provided us with our rugby pitch.

The duty day started at 06.00 in summer – about April to October – and ended at 13.00 hours. After lunch most of us retired for a 'siesta' until about 16.00 hours, when our bearer brought tea. In the winter, which was just like a good British summer, the day started at 08.00 and finished at 16.30. In the evening we played tennis, squash, cricket, football. In summer the temperature sometime rose to 116°F, falling to the eighties at night.

My batman, Umar Khan (in India a batman is called a 'Bearer') was a wonderful chap. He did virtually everything for me. One evening I was due to play squash. My opponent didn't turn up. 'I play with you, Sahib' he said. He took my spare racquet, and over to the court we went, he attired in his normal pantaloons and turban. He beat me comparatively easily. Years later I went out to India again – to the 50th anniversary of the Indian Parachute Brigade – with which my wing had operated. There was the usual squash court attached; in the evening I went round hoping to get a game. To my surprise I found that the pro or 'Marker' as he is called in India – was Hashim Khan, the then World Professional Squash Champion! I asked him where he came from. 'Peshawar,' he replied. I said, 'I wonder if you ever knew Umar Khan, who was my bearer and a very good squash player?' 'He my Uncle' he replied. 'He teach me to play!'

Our duties frequently involved flying over tribal territory. There was thus always the risk that one's engine would fail – or indeed that one might be shot down! In such event, one never knew quite what the reactions of the tribesmen might be. Indeed, they had a propensity for removing one's 'private parts'! Against such an unpleasant contingency, one always carried what was called a 'Ghoulie' chit ('Ghoulies' being the Hindustani for 'balls'!). The tribesmen's love of such wealth as £15,000 ensured that anyone suffering this misfortune would be safely delivered back!

Domestic services were well catered for. The barber, known as the 'Nappi', came to one's room. A shampoo was always included – very necessary in those dusty conditions. The lather was worked up several times. After the first, he would take a sample and slap it on the edge of the basin. Then he would work up a second lather, and slap a second sample beside the first. There was a noticeable difference in colour. Then he would work up a third, and repeat the procedure till he could see no difference in the 'whiteness'.

One's footwear was made by the 'Muchi'. He would copy any style faithfully, and turned out excellent work. Tailoring was done by the 'Dhersi'. Again, he would come to your room, measure you up, and copy any style 'to a T'. I remember asking him to copy an old sports jacket that I was rather fond of. He did – even to the patch on one of the sleeves!

Our main role was guarding the frontier from possible attack from tribesmen, and keeping the peace in the tribal area between British India and Afghanistan. Though it was not mentioned much, if Russia had wished to invade, it would have been through the Khyber Pass.

The main tribes were Mohmands, Afridis, and Wazirs of

Moslem faith. Many had come over the frontier and indeed were free to do so. My bearer was an Afridi.

The British had established strategic forts in the tribal territory. One such was Miramshah, in Waziristan. It was a large fortress, of the 'Beau Geste' variety, surrounded by high walls with towers at the corners. It was manned by an elite force called the Tochi Scouts. The officers were seconded from the Indian Army for two years. There was great competition to get in. The troops were enlisted from the local tribes. The squadron kept one flight there, rotating every two months, and in liaison with Kohat.

The job of the Tochi Scouts was to keep the peace in their portion of tribal territory. Frequently they went out on patrols, known as 'ghashts'. These could be of two or three days duration. They would draw tracks of 20-25 miles per day on the map, finishing back at Miramshah. There was always the chance of hostile activity, and Scouts were sent ahead to cover the flanks and report. Generally the tribesmen were friendly; they realized that the presence of the Scouts deterred migrant tribesmen from attacking their villages. At the end of each day the Scouts selected a suitable place to bed down for the night, scooping a hollow for their bottoms, and wrapping themselves in a blanket. At some points there were small forts; Dosulli comes to mind, and Rasmak – at 6,000 feet above sea level; it was pleasant in summer, but very cold in winter.

The RAF had their own section of the fort at Miramshah. It was separated from the Tochis by a wall, with steps up one side and down the other. We were great friends, and most evenings there was a regular two-way traffic over the wall!

The Tochis used to invite us to go out with them on 'ghasht'. I went on several. It was really tough going, and after three days during which we would have covered sixty-odd miles over rough country, I recall staggering back the last mile or so to a very welcome hot bath and ample 'refreshment'. Sometimes the flight would ascertain where the Tochis were spending the night, and drop food by parachute. The Scouts greatly appreciated receiving a hot curry and (well-packed) bottles of beer. This must have been among the first supply-dropping operations – the forerunner of many I was to participate in some twelve years later in South-East Asia.

Back to life in Peshawar. Apart from our constant training in the various duties of the Army Co-operation pilot, there was plenty to do. Virtually every sport was catered for. Within the squadron there were inter-section soccer leagues, tennis, cricket, squash, rugby and athletics. In the outer sphere of the Peshawar District, there was similar competition with army units. Hunting and polo were also available. We were next door to an Indian cavalry regiment. They laid on courses and offered us the use of their horses (ponies). I recall that we started without a saddle, just walking, with an instructor holding the rein. Then it was 'trrrrot' and so on. All very thorough. Some of our chaps got quite hooked, and joined the famous Peshawar Vale Hunt, the 'PVH'. I tried my hand at polo, but not very successfully. One of our number used to go down to the aerodrome before breakfast, and practice, secretly. It was only when he presented himself for a practice game that it was pointed out to him that you hit the ball with the flat side of the mallet, not the end as in croquet!

There was an interesting social life in Peshawar, largely centred on the Peshawar Club. It was for officers only, together with a smattering of civilians – to wit, the manager of Grindley's Bank (a chap to keep in with!) and the Governor and his staff of civil servants. I recall that the Army predominated. Army officers were still recruited largely from the privileged ranks of society, through Sandhurst and Woolwich. A private income was almost a necessity, as army pay was not very good. The RAF was considered a little down the social scale, being largely short-service entry, with only a handful of officers trained at Cranwell. There was a well-defined 'season', from about October to April, when the weather was cool. Married officers brought their eligible daughters out from Home. They were known as 'the Fishing Fleet'. But there was always a preponderance of males, which meant that even the plainest girls did not lack for 'suitors'. There was always a dance on the Saturday night, at which 'tails' were the order of the day. The girls had programmes and the unattached male would be lucky to win more than two or three dances in an evening. But there was always the bar...!

Towards the end of April it began to get hot. The female population departed for the cooler hill-stations – of Murree and Kashmir...Thus for a time the position was reversed. There were more females than males – until the 'leave' season began.

Leave was very liberal. One's entitlement was two months per year. But one could get 'ten days casual' if one's CO and the exigencies of duty permitted. We were well served by the proximity of Murree, a hill station situated at 6,000-7,000 feet above sea level no more than two hours journey away. It had been developed largely as a military centre to which units could be sent for a month or so to 'cool off' whilst still performing certain forms of normal training. But Kashmir was the far superior place. It catered for most tastes – trekking in the Himalayas, climbing, shooting, living on a houseboat on one of the network of lakes, which included the Shalimar Bagh, immortalized by Amy Woodford-Finden's song 'Pale hands I love...' Some people did a bit of each activity – trekking for a couple of weeks, and then relaxing on a houseboat. Normally two or more would share a boat; there was much haggling about the cost, but, as I recall, one could get a good boat for about 400 rupees a month. At one-shilling-and-sixpence to the rupee, this was £30! With the boat went the staff – usually the whole family, who lived on a smaller boat nearby. The normal method of travel was by shikara, a boat very similar to the Venetian gondola but in shape more like a large punt. It seated two or three in cushioned comfort, with curtains and cover. The crew paddled from behind, often punctuating their paddling rhythm with dirges. One simple one which I remember was 'Whisky...Soda...Pani [water]...Na mankta [don't want]'. The shikaras had interesting names, I remember one – 'Lover Come Back to Me'.

The capital, Srinagar, was a very interesting town, producing rugs, brassware, china and many other articles. It was famous for wood carving; I still have boxes with 20 Squadron crest skilfully carved on them. There was of course the traditional club, the Srinagar Club. It was situated on the Jhelum River, which was connected to the network of lakes. Thus one's communication was by shikara, 'door to door'. On a Saturday night there was always a dinner-dance. 'White tie and tails' were the order of the day. The latter were apt to get a bit creased, particularly on the homeward journey. Falling into the water was not unknown!

But back to squadron life in Peshawar. In 1933 trouble broke out in the Mohmand territory. The Peshawar Brigade, under Brigadier Claude Auchinleck, went into action. No. 20 Squadron started to carry out the duties for which they had trained – mainly reconnaissance, but also some photography. I was out on an early morning sortie on 15 September 1933. We had noticed some activity on a ridge near Yuseph Khel and encountered some rifle fire. As I was about to report it, my engine stopped. I was at about 1,500 feet. There was little chance of landing in one piece, so I chose a spot near the dried-up riverbed. I carried out the forced landing procedure which was a part of our training. The ground was very rough. I held off longer than normal, and eventually 'pancaked' onto the ground. As was inevitable, the undercarriage collapsed, and we came to a stop very quickly. Fortunately we did not turn over. My observer, LAC Skinner, and I got out unhurt. I told him to remove the rear Lewis gun from the scarf-ring, and mount it on a pile of stones, facing the ridge where the tribesmen were assembled. I estimated that our nearest forces were some four to five miles away – beyond the riverbed. There was about half a mile of clear ground over which the tribesmen would have to advance to capture us. Behind us it was about a quarter of a mile to the riverbed. Our instructions were that in the event of a forced landing such as this, we should stay by the aeroplane, to which rescue parties would be directed.

I decided that we should stay until tribesmen came into view, and then we would 'leg it' to the riverbed. Desultory shots were now arriving around the wrecked aeroplane, behind which we were taking shelter, with our Lewis gun covering the open stretch towards the foothills. We had tried to remove the fixed Vickers gun which fired through the propeller, but we did not have the right tools. We therefore removed the lock and the bolt.

We must have been there about an hour when we heard a shout from the direction of the riverbed. Two or three figures had emerged and were gesticulating and shouting to us to join them. Thankfully we shouldered the Lewis gun and a belt of ammunition, and ran to the riverbed – dropping the six feet or so into comparative safety. Our rescuers were a platoon of Sikhs under the command of Captain Gordon, whom I had met in Peshawar. We followed the riverbed for about an hour, and arrived at the Brigade HQ camp, where we were immediately taken to Brigadier Auchinleck's tent for a full debriefing. We had to stay the night – rather uncomfortably – and the next morning were taken back to Peshawar.

There we were immediately seen by our CO, Squadron Leader

Hollinghurst. After listening thoughtfully to our account, he asked, 'Did you switch over to reserve fuel at the correct time?'

That would have been about the time we force-landed! I had to say I didn't think I had.

'So that could have accounted for the engine failure?'

'Yes, I suppose so, Sir.'

Gone was the more romantic reason that I had been shot down!

I should mention here that capture would not necessarily have been the end. We always carried with us a ransom note, which promised to pay £15,000 for the safe return of a prisoner. It was vulgarly known as a 'ghooli chit' ('ghoolies' being the Hindustani for 'balls') owing to the propensity of tribesmen to remove these rather personal parts. The commander of the Sikh platoon, Captain Gordon, was given the immediate award of the Military Cross, for 'rescuing an RAF pilot and his observer, who would otherwise undoubtedly have fallen into the hands of the enemy'.

These operations were later recognized as a campaign justifying the award of the Indian General Service Medal with the clasp reading 'Mohmand 1933'.

My tour of duty – and virtually my short service commission – ended in February 1935, and I was posted back to the United Kingdom. I had quite a job packing the souvenirs and rugs I had acquired. Most of them, of course, were assigned to Cox's and King's for forwarding. My journey home was a repeat of the journey out – Frontier Mail to Karachi, then SS *Nervassa* for the two-to-three week voyage to the UK. After a few weeks leave, I reported to Uxbridge, where five years earlier I had started my RAF service. This time it was to be discharged. After much signing of papers and handing in of equipment, I was given a farewell handshake and a cheque for £260 – my gratuity. While strolling through the town, I had seen a Triumph Nine two-seater with dickey. It rather took my eye. It was a 1933 model – price £160. I went into the garage, presented my cheque, and became its proud owner.

Thus on 12 April 1935 I found myself with a car and a few pounds in my pocket, but no job. I returned to my home, to be welcomed like the Prodigal Son by my ever-loving parents.

CHAPTER V

Civilian

Sywell & Perth

At this time momentous events were beginning to take place in Europe. Hitler was on the move. Belatedly our government were realizing that war was inevitable – and that we were ill-prepared, particularly in aircrew and aircraft for the RAF. A scheme was introduced whereby pilots' initial training was handed over to Civil Flying Schools. The '*ab initio*' course, which had taken six months in the RAF, was reduced to two months! Entrants virtually remained 'on probation' until they had passed the course. Those who did not pass were thanked, and went their ways. No uniform or unnecessary equipment was issued.

Our Short Service Commissions included a commitment to four years on the Reserve of Air Force Officers (RAFO). This entailed reporting to a Civil Flying School for a fortnight every year, and 'refreshing' oneself – usually on a Tiger Moth, Avro-Tutor or similar machine. It was a 'jolly', and one spent a lot of the time visiting old friends at other aerodromes. I had only been a few weeks at home, looking for and writing about jobs, when I received an OHMS letter. It informed me that I should go almost immediately for my first annual Reserve training, and that it could take the form of a Flying Instructor's Course; there was a place available at No. 8 Reserve Flying Training School, Sywell, Northants in June (1935). The normal flying instructor's course was carried out at the prestigious Central Flying School – and took two months. I was intrigued to see how it could be crammed into two weeks at a civil school. I duly reported, and was assigned to Pilot Officer Goldsmith, recently commissioned.

The school was run on contract, by the well-known firm Brooklands Aviation. They had also taken over the Northampton Flying Club, which shared the aerodrome. I duly completed the course, having got on well with Goldsmith, and took my test with an independent examiner appointed by the Guild of Air Pilots. I passed, and this entitled me to a Flying Instructors' rating on my 'B' Professional Pilot's licence, which I had acquired recently.

I was packing up ready to leave Sywell, when I was told to report, to the Chief, Wing Commander I.C.W. Mackenzie. He said, 'Have you got a job to go back to, Don?' (my other name!)

'No, sir,' I replied.

'How would you like to take over as Club Instructor? Tommy Rose is leaving.'

'I am very honoured; I would like to,' I said.

So I became the instructor of the Northants Flying Club. Tommy Rose, from whom I was taking over, had been offered a job – by, of all firms, Seager's Gin! Tommy was a great character – DFC in the Royal Flying Corps etc. He had achieved quite a name flying to the Cape, and participating in a race to Australia. Later he was to become Chief Test Pilot of Reid and Sigrist. But now his new job was to sell Seager's Gin! They gave him a small aeroplane – a Puss Moth, I think – and he had to tour all the aerodromes, introduce himself at the bar, and dispense free gin to customers! Our paths were to cross again, and we became great friends.

The job of a Club Flying Instructor, at the average club, was more than just flying. He was also the social host – and even had to help out behind the bar! My salary was £5 per week, plus 'flying pay' – of four shillings an hour. I suppose this brought it up to about £10 per week. One day one of our members, Bill Abbot, whose name is well-known in the boot and shoe trade, for which Northampton is renowned, came in escorting a very lovely girl. He introduced me: 'Mollie, this is George Donaldson, our instructor – Miss Mollie Brown'.

'Are you flying this afternoon, Bill?' I asked.

'I don't think so; we'll just have a drink,' he replied.

Now it so happened that the club had been 'loaned' a new aeroplane – the 'BA Swallow'. The firm, British Aircraft, of Hounslow, had said, 'Give members who might be interested a free flight.' I asked Bill if he would like one. Miss Brown said, 'Oh, I would; could I come too?'

Alas it was only a two-seater. On an inspiration I said, 'I'm afraid I can't give two flights – would you like to toss for it?'

'OK,' said Bill. Miss Brown won the toss; we duly strapped her in, and off we went. Through the intercom I asked her if she would like to loop the loop.

'Ye...es,' she replied.

We looped the loop. I obtained her telephone number, and invited her to a tramp's party at the club. She accepted; we fell in love.

One of the old friends I met at Sywell was Jimmy Sholto-Douglas. He had served with me in No. 20 Squadron on the North-West Frontier of India, and was now an instructor at the school. One day he came to me and said, 'Have you heard that the Navy are looking for pilots?' 'No,' I said. The Navy had long been trying to get control of their flying services. Up until 1918 they had had their own service, the Royal Naval Air Service. It had amalgamated with the Royal Flying Corps to form the Royal Air Force in 1918. Since then the RAF had carried out all the Navy's requirements. At last the Government had agreed that they should again have their own aircraft. The new force was to be called 'The

Fleet Air Arm'. Their problem was to find the pilots. An obvious source of experienced pilots lay in the many Short Service Officers which the RAF had released over the years, and whom the Air Ministry said they would be prepared to release from their Reserve commitments. The Admiralty accepted the offer, and invited applications. Jimmy applied and was accepted. I thought about it, but did not apply. I have sometimes wondered if I 'missed a turning'. Jimmy was to do well. He became a commander and served in the carrier HMS *Courageous*, earning a DSO.

One day the phone rang. 'This is John Paget speaking; could you fly me up to Liverpool urgently? I am due to speak at the election meeting of Randolph Churchill.'

I had no appointments, so I said 'Yes.'

Right,' he said, 'I'll be with you in half an hour.'

I told Jim, our ground engineer, to get the club aircraft ready. John Paget duly arrived, with quite a retinue. I indicated the 'Moth', and said he would need to wear a flying-suit and helmet.

'Good Lord,' he said, 'we're not flying in this are we?'

'Yes, I'm afraid so,' I replied.

'But there's only room for one – what about my Chairman?'

'Sorry,' I said, 'it's the only aircraft available.'

After a pause he said, 'Ah well – I've got to go.'

We dressed him in the flying-suit, strapped him in, showed him how to work the primitive intercom, and off we went.

After we had been flying for about twenty minutes, he shouted, 'Hello – can you hear me?'

'Yes,' I replied.

'I've got it', he said, 'you shall be my Chairman.'

'Not likely,' I replied.

We duly landed at Speke, to be met by a car and whisked off to the Adelphi Hotel. I was introduced to Randolph, and during a hasty meal they again tried to persuade me to act as Chairman. I agreed, subject to being told what to do. 'Oh, you'll be given a gavel: if the meeting gets too rowdy, you bang the table and shout, "Order, order!"' The meeting turned out to be *very* rowdy – and I had occasion to bang the table several times.

The next morning we flew back to Northampton. When we were about six miles from the aerodrome John shouted, 'There's my house down there; could you land near it?' I circled, saw a field that looked OK, and landed. He waved me off, and later rang expressing appreciation, and saying that he would join the club and learn to fly.

I did not remain a club instructor at Sywell for long. Civil Flying Training Schools were starting up at many places. It was becoming a flying instructors' market. I applied to Reid and Sigrist, who were opening a school at Desford, offering – as I recall – '£400 a year plus so much per flying hour'. I applied to Airwork Ltd, of Heston, who had been awarded a school at Perth. They were offering £700 a year flat. I was subjected to a flying test and offered a post. I accepted.

Perth

My first instructions from Airwork were to report to De Haviland's at Hatfield, collect a brand new Tiger Moth, and fly it up to Perth. On 6 January 1936 I reported to Hatfield, there to meet seven or eight of my colleagues-to-be, assembled for the same task. I knew one or two of them; we were to become a close-knit community. It was a severe winter; snow had been forecast for the whole of the country. We had no flying suits, and Hatfield could not help. So we put on all the clothes we could, but nevertheless were not looking forward to some five hours flying in open cockpits!

The next morning we took off in loose formation. As we flew north, the snow increased. We landed after three hours at

Doncaster, refuelled and flew on to Newcastle. Here we night-stopped. The next morning we were off again, to arrive at Perth after one hour and forty minutes. The whole of Scotland was white with snow and we had difficulty identifying the aerodrome, which was at Scone, just a few miles outside Perth. Eventually someone saw a Very light, and we saw a 'T' laid out on a snow-covered area. We landed and were met by a wonderful chap – John (later Sir John) Primrose. He was a councillor, and it was almost entirely his efforts which had resulted in Perth building one of the first municipal aerodromes. He was standing by a large brazier with a bottle of whisky in his hand, and glasses on a table.

'Come away, boys; ye'll be needing this,' he said.

Numb with cold as we were, he couldn't have spoken a truer word.

That same night we were on the 'Night Scot' back to London, and then out to Hatfield to collect more aircraft. I made three such trips.

No. 1 Course, of No. 11 Elementary Flying School, arrived. The snow melted, leaving the grass airfield a sea of mud. We were told to move lock, stock and barrel – to Abbotsinch, an RAF airfield just outside Glasgow. It was then solely occupied by the City of Glasgow Auxiliary Squadron. This airfield was just about a mile or so from the River Clyde, and John Brown's famous shipyard. On the stocks dominating the skyline was the *Queen Mary*. In the ensuing weeks we were very thankful for her as a landmark enabling pupils to find the aerodrome. One day, it had disappeared! It had been launched. It was surprising how many pupils got lost. A phone call would come, as one was anxiously awaiting one's pupil who had been sent off forty minutes before. 'Sorry, sir; I got lost, but I found Prestwick, and have landed safely here.' So off to Prestwick to collect him; and be glad that he had got down in one piece!

At that time Prestwick was the home of 12 EFTS, who were its first tenants. It was run by Scottish Aviation, under a Wing Commander McIntyre. He had achieved some fame by leading an expedition, with two or three Westland Wapiti aircraft, to fly over Mount Everest. His exploits are well told in a book *First Over Everest*. Prestwick was to become famed for its freedom from fog. As a result it became the main entrepot for aircraft which were to be supplied in increasing numbers by America, under, initially, the 'Lend-Lease' agreement.

After a month or so, Perth was declared fit, and we returned. We flew hard, turning out pilots at the rate of about thirty every two months. Some of them were to achieve distinction; many of them, alas, were to be killed in a war which was now inevitable. But the life at Perth was pleasant. The 'Natives' were friendly, offering us the pursuits such as fishing, shooting, golf, for which it is famed. I soon decided that it would be more pleasant to live away from the aerodrome, so I visited a house agent in Perth to see what might be available. He showed me a flat in Rose Terrace in an old Victorian house, facing the North Inch and the River Tay. The North Inch was a historic area of green. In 1396, in the presence of King Robert III, the 'Battle of the Clans' had been fought there by selected warriors of the Mackintosh Shaws and Cummings. Its object was to settle once and for all the age-long feud which had existed between them. The Cummings were defeated, and the remnants of their force plunged into the River Tay and escaped.

Among my colleagues was one, George Walker, with whom I got on well. I asked him if he would like to share a flat with me. George was a canny Scot, about eight years older than I, who had been in the Royal Flying Corps. He said, 'Give me a day or two to think about it. The next day he said he would like to. The rent of the flat was £7.10 per month! When we had settled in, we said,

'Why don't we engage a housekeeper/manservant, and live in some style?'

The depot of the Black Watch Regiment was just around the corner, so I went round and asked if I could have a word with the Adjutant. They were most courteous, and ushered me in. I explained our position, and asked him if by any chance they ever had mess stewards or batmen due for release, who would like a job. He made enquiries and said, 'Well, it so happens we do. A mess steward, name of Hart, is due for release in a few days.' We interviewed him – and engaged him. We fitted him out with a white jacket and a short monkey jacket with brass buttons, for 'occasions'. He wasn't a good cook, but otherwise he was a great success.

Perthshire Cricket and Rugby teams played on the North Inch, almost opposite our flat. I was privileged to play for the latter. It was quite an experience to be able to change in one's own home, trot out onto the field, and go straight back into one's own hot bath!

Living in Perth had one penalty, albeit quite an enjoyable one. Most of our friends used to come into the city in the evenings to have a few drinks. Our favourite place was the Sundown Bar at the Royal George. In those days closing time in much of Scotland was nine o'clock. Inevitably, after closing time the suggestion would be 'Let's go round to Don and George's flat.'

Since leaving Northampton I had kept in close touch with dear Sue. She used to come up for holidays, often driving up in her little Austin Seven – which she had won in a raffle! One of my friends, Robin Rendle, lived with his mother nearby. She always welcomed Sue and looked after her like a mother. One of our favourite jaunts was to dine and dance at Gleneagles on a Saturday night. It was about fifteen miles from Perth. On one occasion Sue's visit coincided with a long-standing invitation I had, to act as

44

usher at an old friend's sister's wedding in Liverpool. I was reluctant to spoil our dinner and dance at Gleneagles, so on the Saturday morning I borrowed one of the School's Tiger Moths and flew down to Speke, Liverpool Airport, the flight taking about three and a half hours. I was met at Speke by one of my friends with a hired morning suit into which I changed. After the ceremony, the reception was held at a country manor, over in Wirral. Time would never have permitted me to attend the reception travelling in the normal manner, involving as it did crossing and recrossing the River Mersey. I knew the Wirral countryside well, and was fairly confident that I would find a field near the manor in which I could land my Tiger Moth. I did – and attended the reception, drinking the health of the bride and groom. Some of my friends came to see me off on my return journey to Perth. It was only after I'd taken off that I realized I was still attired in the hired morning suit! I landed back at Perth around six-thirty, changed into a dinner jacket, and drove over to Gleneagles. Sue had been taken over by my friends and was already thoroughly enjoying herself. I joined in and we had a wonderful evening. At midnight we stood to the strains of the National Anthem, then drove the fifteen miles back to Perth. It had been quite a day.

Perth aerodrome was devoted almost entirely to the Flying School and the local Flying Club, but just occasionally a charter aircraft would call in. One afternoon, as we were finishing flying, an aircraft landed, a passenger alighted and was taken away in a waiting car. The pilot had apparently decided to stay the night and was putting his aircraft into the hangar. Later he came over and asked where he could phone for a taxi. 'If you're going into Perth I can give you a lift,' I said. He introduced himself as Bill Leadley of Personal Air Services, Croydon. It was one of the smaller charter companies which sprang up between the wars. I said, 'I have a spare bed – you are welcome to stay the night with me.' He accepted gratefully. The next morning I took him up to the aerodrome and saw him off. A few weeks later another charter aeroplane landed and taxied in. The pilot got out, and I saw one of the ground crew pointing in my direction and gesticulating that he wanted to see me. He introduced himself as Joe Birkett, of Birkett Air Services, and said, 'Could we have a chat somewhere?'

I took him over to the clubhouse and ordered coffee.

He said, 'I am in partnership with Brian Allen. We have won a contract to open an RAF Reserve Flying School. A friend of mine, Bill Leadley met you. He thought you might be a good person to run our school – as Manager and Chief Instructor. Would you like the job? Please consider it and let me know as soon as you can. By the way, it is No. 29 (E & R) FTS and will operate at Luton. The salary will be £800 per annum, plus a percentage of the profits.'

I decided to take the job. When I told my colleagues, three of them asked if they could come with me. I was very touched and of course said yes. They were the three best instructors in the school. One, Geoffrey Jarrett, had been out on the North-West Frontier with me in No. 5 Squadron at Quetta. Stuart Miles was an ex-fighter pilot. Nick Nicholson was ex-Bomber Command.

I said goodbye to Perth with sadness. I had enjoyed a good life there. But I was to see it again at a later date.

CHAPTER VI

Back in Uniform

Luton, Sealand, Upavon, Thruxton

No. 29 was one of the schools virtually 'mushrooming' to cope with the entrants to the RAF Volunteer Reserve who were now flooding in. The VR was equivalent to the Territorial Army in that their training would be mostly at weekends and in the evenings. It was therefore important to form training schools near the areas where they lived. We started with six transferred from Hatfield, but soon had more than we could cope with. I sat on the local selection board, chaired by a retired Wing Commander. My contractors showed me round our accommodation. The Ministry had built a hangar and a row of wooden huts. They also provided the aircraft. We started with six Miles Magisters. Brian and Joe said, 'We've got you a chief engineer from BOAC – otherwise it's all yours.' It was quite a challenge recruiting the cross-section of staff required, but we managed, and were soon in action. One of the requirements was that there should be a qualified medical attendant on duty at weekends. Sue had trained as a VAD! I immediately recruited her, and she would drive down the sixty-odd miles in her Austin Seven, attired in her smart VAD uniform – and be with me for the weekend. Needless to say, I had quite a lot of trouble tactfully fending off pupils who, quite under-standably, took a fancy to her!

As the school grew, our aircraft establishment increased to eight, then to twelve Magisters. Then, as reservists completed their 'ab initio' course, we were given Hawker Harts for advanced training. Navigator training was also introduced, for which we were given three Avro Ansons.

But the clouds of war were gathering rapidly. One day having our morning coffee we heard on the radio the fateful words of the Prime Minister Neville Chamberlain:

> This morning the British ambassador in Berlin handed the German Government a final note, stating that, unless we heard from them by eleven o'clock that they were prepared, at once, to withdraw their troops from Poland, a state of war would exist between us. I have to tell you now that no such undertaking has been received, and that consequently this country is at war with Germany.

Our immediate instructions were to don uniform and carry on as usual. Many of our pupils left their civilian jobs and presented themselves for full-time flying. With weekends being our busiest times, I had long been working seven days a week – the 'day' being from 0900 hours until dusk – with night-flying thrown in!

During these first weeks of the War, I was in daily touch with RAF Reserve HQ and my contractors. One day I received orders to move the school lock, stock and barrel to Hanworth, there to amalgamate with No. 4 EFTS, run by Blackburn Aviation. Luton was required for other purposes. There was reorganization – in many cases disorganization – in the flying training world. I decided that this life was no longer for me, and applied to rejoin the RAF proper, going to one of my old squadrons, '2' or '20' if possible. I was informed that my application to get back in to the Service had been approved – but that my experience as a flying instructor was still required. I was posted to No. 5 Service FTS Sealand, as Flight Commander 'C' Flight, with ten or so Airspeed Oxford twin-engined monoplanes. My Chief Flying Instructor was Squadron Leader Laurie Sinclair, whom I had met before, and who was to achieve distinction.

I was not at 5 FTS long. After about six months – in May 1940 – I was posted to the Central Flying School as an instructor – our job, to train selected pilots to become flying instructors. It was regarded as a prestige posting. The Central Flying School had been formed by the legendary Smith-Barry to carry out the 'Sequence of Instruction' which he initiated. Its role was now to train potential flying instructors, developing standards in line with advances in aircraft design. I was Flight Commander 'D' Flight. Our CFI was Wing Commander A.J. Holmes. He was a good chap – a noted cricketer, having played for Sussex and captained an England touring team. Another colleague worthy of mention was Arthur Donaldson, who commanded 'C' Flight. He was one of three brothers, the others being Teddy and Baldy, who were to achieve fame. We tried to trace a family connection but there was no obvious one.

My first course was an unusual one. It comprised a mixture of older pilots, many of whom flew in the Royal Flying Corps in the First World War, or had done a little club flying, or flown with the Civil Air Guard. The idea was to 'vet' them, and if found suitable, give them refresher flying and train them as '*ab initio*' flying instructors. On reading the names I saw 'Smith-Barry'. Surely it couldn't be *the* Smith-Barry, the founder of the Central Flying School some thirty years ago, and initiator of the 'Sequence of Instruction' which had changed little over the years? It was. I took him on as my personal pupil. He had not flown much since the First World War, so needed a little practice before starting on the instructors' 'patter' and demonstrations. When we did, I found that he was very prone to deviate, and 'do it his way'. Pupil-instructors were encouraged to adapt the 'patter' to their own style, but he had his own ideas about the flying demonstrations.

For instance, instead of putting the nose of the aircraft down to gain a little extra speed in a gliding turn, he pulled it up! When asked why, he said, 'It gives me a better chance to look around!' Altogether I wasn't too happy about him, as I also heard that he might 'pull strings' and get command of his own school. I discussed the case with my Chief Instructor. He said 'I'd better check him, but good Lord, we can't fail the chap who founded the Central Flying School!' He checked him – and reluctantly agreed with me. So I achieved the doubtful distinction of failing the founder of the Central Flying School! He became a ferry pilot, and not long afterwards had an accident with an Airspeed Oxford. He then got a ground job as an adjutant, I think.

Another course comprised civilian pilots who had volunteered to join the Air Transport Auxiliary ('ATA', as it became known). The object of their course was to familiarize them with RAF aircraft, which differed in certain respects from the civil. Among that course were a number of well-known names in Civil Aviation, including Amy Johnson, who had quite recently completed her epic flight to Australia in a DH Moth!

Again, I was not to be at Upavon long. The demand for pilots – and, ergo, for flying instructors – was now so pressing that the Central Flying School could not cope. Additional 'Flying Instructor Schools' were therefore formed. One such – No. 5 Supplementary Flying Instructors' School – was to be stationed at Perth. I was posted to command it. The date – November 1940. The post carried the rank of Squadron Leader (Acting) to which I was promoted.

Our 'pupils' were an interesting lot. A large proportion of the initial intake were again ex-First World War RFC pilots who had not kept up flying after their war. They needed a bit of refresher flying first, but then coped pretty well. Among them was one called Donald. Then, believe it or not, a new arrival was ushered into my office, and introduced as 'Sergeant Don'! After a few preliminary remarks, he said, 'I think I'd better come clean with you, Sir. After the War [1914-18] I stayed on in the newly-formed Royal Air Force. I became a Flight Commander [later changed to Flight Lieutenant] and was a member of the King's Flight at Hendon. One day the Prince of Wales asked to see me, and said, "I would like you to teach me to fly, Don." I said, "I would be greatly honoured, Sir." He was an apt pupil. There was some concern in the higher echelons about his flying solo, but he overcame them and succeeded in reaching "Wings" standard. I was drinking rather a lot at this time, and put up one or two "blacks". I was dismissed the Service. But I have managed to get back – as a non-commissioned officer.' I took Don on as my personal pupil. He was very good, and for the rest of the War did a very useful job training *ab initio* pilots. Among other pupils who passed through, I recall Flight Lieutenant Skinner, founder of the well-known London jewellers of that name, and Sergeant Trueman, whose sister Christine Trueman a well-known tennis player. I see from my log-book that I went over to Abbotsinch where there was a Spitfire OTU, and persuaded them to lend me one in which I flew for forty-five minutes. It was a wonderful experience.

During all this time I had been seeing my dear Sue as often as the conditions of service allowed. Early in 1941 she came up to Perth for a few days. I said. 'Let's get married.' She agreed. We went over to Edinburgh, and were married in a registry office on 20 February 1941. We always remembered the Registrar's words as he shook hands. 'Very short – but very binding.' I had only felt able to take forty-eight hours leave, so, after a one-night 'honeymoon' in Edinburgh, we returned to Perth – and duties.

We were fortunate to know the Secretary of the Flying Club, who was also a solicitor. I asked him if he had any small properties on his books. 'We have a small cottage, once the gardener's, at the bottom of our garden. Would you like it?' We saw it, and fell in love with it. 'How much?' I asked. '£200,' he said. We bought it. But our stay was not to be long. One day I received another phone call: 'You're posted to No. 5 (P)AFU Hullavington as Chief Flying Instructor.' (P)AFU was Pilot's Advanced Flying Unit. Hullavington was a large unit, comprising three flights of Miles Master aircraft and Hurricanes, for the training of potential fighter pilots, and three flights of Airspeed Oxfords for those likely to go on bomber and other duties.

We said farewell to Perth for the second time – and to our dear little newly-acquired home – and moved to Hullavington. The move carried with it promotion to Wing Commander (Acting). The date, July 1941.

Hullavington was a continuation of the pilot training ritual. Though I realized its importance, I was becoming increasingly frustrated with it. I considered I had done my whack. I had applied to be posted to an operational unit whilst at Perth. The application was refused. I applied again. I was again turned down. The formula was always, 'Until a suitable replacement can be found . . .'

Meanwhile dear Sue and I found a furnished cottage nearby, and were very happy. Often, when I was night-flying, she would come with me, and sit in the Mess till I had finished. We were able to get the occasional day off and visit Bath and other interesting places nearby.

I spent most of my time checking out pupils on Miles Masters and Airspeed Oxfords. Miles Master pilots graduated to the Hawker Hurricane. I used to enjoy flying it – a lovely aeroplane. Such was the volume of flying that most Training Command stations had satellite airfields. One of ours was adjacent to the beautiful village of Castle Combe.

But again, it was not to be for long. One morning I received a

From: A/S/Ldr. G.F.K. Donaldson (33068),
      No. 5 .F.I.S. PERTH.

To: Air Officer Commanding,
      No. 51 Group.

Date: 16th May, 1941.

Subject: P O S T I N G.

Sir,

          Further to my conversation with you,
I have the honour to request that I may be posted
to an operational unit.

          During my Royal Air Force service I was
for 3½ years an Army Co-operation pilot. I have
since then made some study of Navigation and hold
a Civil Air Navigator's Licence. I have recently
flown modern types of both twin and single engined
aircraft.

                    I have the honour to be,

                         Sir,

                    Your obedient servant,

                         COPY.

SECRET.   POSTAGRAM.

                              FTC/36872/PLO.

                              23.5.41.

To: H.Q. No.51 Group.

From: H.Q. Flying Training Command.

          F/Lt. (A/S/L) G.F.K. Donaldson.

1.   Reference your postagram 51G/957/350/P.2.

dated 19th May, 1941, it is regretted that until

such time as the C.F.S. can provide a suitable

relief it is necessary for Squadron Leader Donaldson

to remain in his present post.

                    Sgd. ? ? Hail.

                    Squadron Leader.

1st application for posting to an operational unit – May 1941.

53

A/Wing Commander G.F.K. Donaldson (29068)
R.A.F. Station, HULLAVINGTON,
Nr. Chippenham,
Wilts.

11th. November 1941.

Sir,

    I have the honour to request that this my application for posting to an operational unit may be again considered. For the past 6 years I have been engaged continuously in R.A.F. Instructional Flying as follows:-

Jan. 1936	–	July 1938.	Instructor	No. 11 E.F.T.S.	2118.05 (S.S.)
July 1938	–	Oct. 1939.	C.F.I.	No. 29 (SWIFTS	302.00 (S.T.)
Nov. 1939	–	May 1940.	Flt. Comdr.	No. 5 S.F.T.S.	179.55 (T.E.)
May 1940	–	Nov. 1940.	" "	C.F.S.	274.30 (T.E.)
Nov. 1940	–	July 1941.	D.F.I.	No. 3 S.F.I.S.	263.30 (.E.)
July 1941	–	Oct. 1941.	C.I.	No. 9 S.F.T.S.	136.15 (S.E.)
					3274.15

    I now find myself suffering from an inferiority complex when among my contemporaries and even juniors in Training Command, and feel that I could do a better job in an operational unit.

    If my application is favourably considered, I am very keen to get on to Flying boats, having made a study of Navigation and being a qualified Air Navigator.

I have the honour to be,
Sir,
Your obedient Servant,

The Officer Commanding,
No. 9 Service Flying Training School,
R.A.F. HULLAVINGTON, Nr. Chippenham,
Wilts.

2nd application for posting to an operational unit – Nov 1941.

call from HQ telling me that I was to he posted, again, to the Central Flying School, Upavon – this time as Deputy Chief Flying Instructor. 'Speedy' Holmes, who had been my CFI, was now the Station Commander. It was the same training routine, this time checking out flying instructors. The Station Commander's official 'quarter' had long been a lovely old house in the village of Upavon, called The Beeches. 'Speedy' did not want it, so he offered it to Sue and me. We accepted gratefully.

    It was on the site of an old monastery and was alleged to be

TELEPHONE: TERminus 3366.
~~HOLBORN 3311~~
Extn. 4157.

Any communications on the
subject of this letter should
be addressed to :—
THE UNDER SECRETARY
OF STATE.

and the following number
quoted :—

Ref.No.

AIR MINISTRY,

~~LONDON,~~ ~~W.C.2.~~

CZECHOSLOVAK INSPECTORATE,

19-29,Woburn Place,

London,W.C.1.

September 20th, 1941.

It gives me great pleasure to award you
THE CZECHOSLOVAK AIR FORCE PILOT'S BADGE, as a mark of
gratitude and appreciation of all the help and co-operation
you have given to the members of the Czechoslovak Air
Force serving with the R.A.F.V.R.

Air Vice-Marshal K. JANOUŠEK,
Inspector General of the C.A.F.

W/Cdr. G.F.K. DONALDSON,
9.S.F.T.S., R.A.F.,
HULLAVINGTON, Wilts.

REPUBLIKA ČESKOSLOVENSKÁ.
MINISTERSTVO NÁRODNÍ OBRANY.

Ministr národní obrany

uděluje p. W/C. G.F.K. D O N A L D S O N

čestný odznak československého letectva.

Tento průkaz opravňuje majitele k nošení čs. pilotního odznaku.

Evid. čís. 189/1941.

Ministr národní obrany:

Award of Czechoslovak Air Force pilot's badge.

55

Wing Commander G. F. K. Donaldson,
No. 9 Service Flying Training School,
HULLAVINGTON.

29th. September 1941.

Sir,

I beg to express my appreciation of the great honour you
have conferred on me in awarding me the Pilots' badge of the
Czechoslovak Air Force.

Anything I have been able to do for your countrymen has
been a pleasure, and was done in the spirit of mutual co-operation
which will, ere long, bring victory to our cause.

The Inspector General,
Czechoslovak Air Force,
19-29, Woburn Place,
LONDON. W.C.1.

Letter of thanks for the award of the Czech Air Force pilot's badge.

haunted by a monk. One of my predecessors, D'Arcy Greig, who
achieved fame as a member of the High Speed Flight which won
the Schneider Trophy – and who held the world speed record for a
time – averred that he had seen it. We had a batman, a wonderful
chap called Glasby. One day Sue and I had been out, and when we
returned we were met by a very excited Glasby. 'I've seen it, Sir –
the ghost. I was walking up the garden and was attracted by a
movement of the bedroom curtains. Then I saw the head and
shoulders of a man in black. I rushed up to the house, and up the

56

stairs: but there was no sign of anyone.' Alas, we never saw him; but sometimes, when I awoke in the night, I had peculiar sensations that there was someone about.

One day Sue said, 'I think I'm going to have a baby!' We were very thrilled, and couldn't get to the doctor quickly enough to confirm – which he did.

On my first day of duty I was arranging my office when there was a tap on the door, and a diminutive airman entered.

'Your coffee, sir,' he said.

I asked him his name.

'648236 AC Richards.' he replied. There was something not very 'Service' about him; perhaps it was his smallness.

'What did you do in civilian life?' I asked.

'I was a jockey.'

'You're not Gordon Richards?'

'No I'm Cliff; Gordon's my brother.'

Of course; Cliff was a very good jockey in his own right. He told me he had volunteered, but did not come up to medical standards. However, he lived nearby, and 'the powers' made an exception. They said he could join and remain always at Upavon. So he became the ACFI's coffee-maker etc.

With the outbreak of war, most well-known sporting activities had ceased. However, the importance of maintaining our racing bloodstock was realized and the two race-meeting venues were kept going – Newmarket and Salisbury, some six or so miles from Upavon. Cliff asked me if he could have permission to ride when there was a meeting at Salisbury. Of course I said 'yes'. Before a meeting he would come in and say, 'I'd put a bit on Fine Fellow [or some such], Sir.'

'How about a pound each way?'

'No, Sir – to win. If it doesn't look like winning it won't be placed!'

Sometimes he'd say, 'Leave it to me, Sir.'

As often as not, he'd put the winnings on my desk with the Monday morning coffee. One day he said, 'Me brother, Gordon, would like you to come out to dinner and bring some of your friends; it'll be at the Savernake Hotel.' I replied, 'Thank your brother, and say we much appreciate the invitation.' I invited a couple of friends and on the evening appointed we went along to be met by Gordon and three or four of his friends, among whom were well-known trainers and a world-famous billiards player called 'Melbourne' Inman. Despite the severe rationing, we had a marvellous dinner – with wines! Obviously Gordon had some 'pull'!

On the Home Front Sue and our first-born-to-be were coming along well. With a week or so to go, there was a small Mess party. Sue would never miss one, but this time she said, 'I don't think I'll come, darling – but you must go.' The Mess was only five minutes away, and Glasby was there, so I went. I hadn't been there half an hour when the Mess Steward approached and whispered, 'You're wanted on the phone – it's your wife.' 'Darling, I think it's coming,' she said. I called the doctor and our nurse, and hurried down. At 10 p.m. on 24 July my darling Sue gave birth to a lovely girl. We called her Sheila.

Only about three miles from Upavon was Netheravon. Traditionally it was the home of No. 2 Flying Training School, but recently it had been taken over by the newly formed No. 38 Wing, comprising three Squadrons, Nos. 295, 296 and 297. This was the start of the Airborne Forces – paratroop-dropping and glider-towing – so successfully pioneered by Germany. I dropped in one day, and learned that the CO was a Group Captain Norman. 'Not Sir Nigel Norman?' I asked. 'Yes,' was the reply. Sir Nigel was a director of Airwork Ltd., who had been our contractors at Perth. I had met him several times and got on well with him. He had been

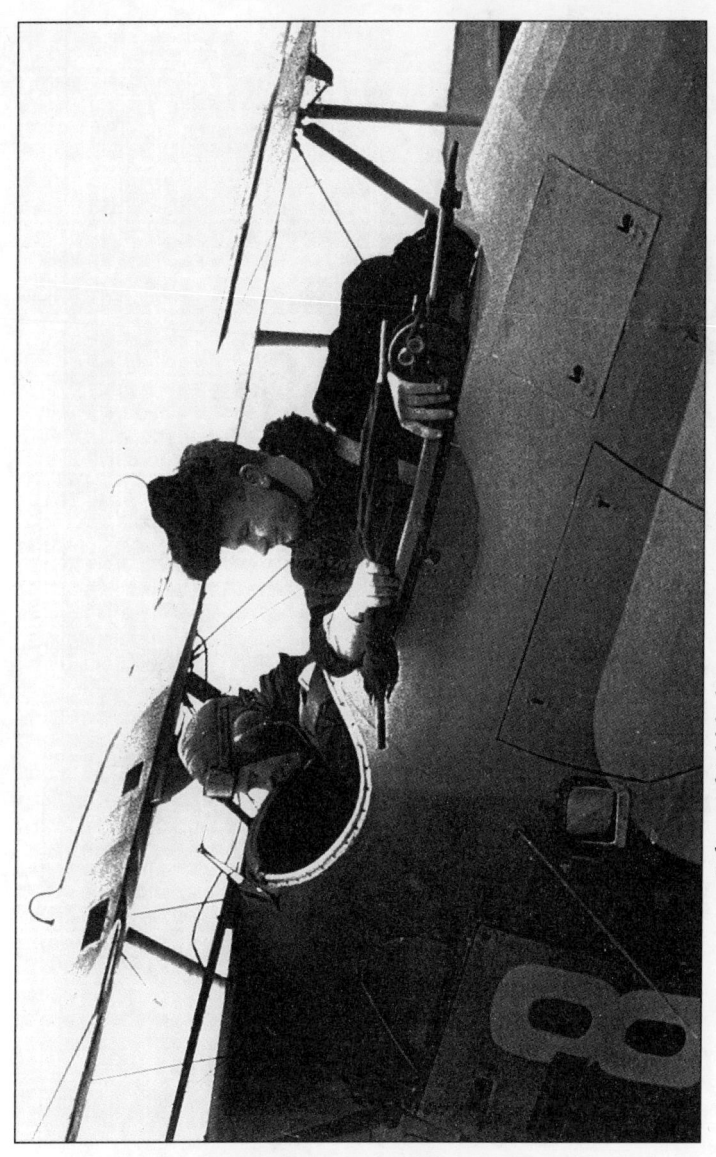

Pilot and 'old lady' passenger – a flying training student.

Hawker 'Hart'.

9th July, 1942.

Sir,

    The King will hold an Investiture at Buckingham Palace on Tuesday, the 28th July, 1942, at which your attendance is requested.

    It is requested that you should be at the Palace not later than 10.15 o'clock a.m.

DRESS-Service Dress, Morning Dress or Civil Defence Uniform.

    This letter should be produced on entering the Palace, as no further card of admission will be issued.

    Two tickets for relations or friends to witness the Investiture may be obtained on application to this Office and you are requested to state your requirements on the form enclosed.

    Please complete the enclosed form and return immediately to the Secretary, Central Chancery of the Orders of Knighthood, St. James's Palace, London, S.W.1.

       I am, Sir,

          Your obedient Servant,

                      Secretary.

Investiture and tea at Buckingham Palace.

a keen Auxiliary Air Force pilot – a member of 602 City of London Squadron. I told him what I had been doing since leaving Perth and of my frustration at being unable to get out of the flying training stranglehold. He said, 'It so happens that I need a CO for 297 Squadron. If you can get your release from CFS, I would be happy to have you take over.' I had not been pulling my weight at CFS. I knew it, and my CO knew it. So I was at last released and took command of No. 297 Squadron, stationed at Thruxton, nearby. I counted myself extremely fortunate.

CHAPTER VII

Action at Last

No. 297 Squadron was one of three – the others being 295
and 296 – which comprised No. 38 (Airborne Forces) Wing.
The Wing was equipped with Armstrong Siddeley 'Whitley'
aircraft, the 'cast-offs' of Bomber Command, which was now re-
equipping with the Halifax, Stirling and Lancaster. We were the
beginnings of the Airborne and Air Supply Forces which were to
play such a vital role in the waging of the War, dropping
paratroops, towing gliders, transporting troops and equipment,
and supplying them. The Army had already formed a Paratroop
Regiment, a Glider Pilot Regiment and Air Supply Units. Our role
was to work with them.

The 'Whitley' carried a 'stick' of ten paratroopers and their equipment. They were dropped through a hole in the floor. Each
parachute's opening mechanism was attached by a static line to an
overhead rail in the aircraft, such that when the paratrooper jumped
out, the line tightened and automatically operated the parachute's
ripcord, opening the parachute. A jump-master was an essential part
of the 'stick'. His duty was to ensure that each parachute was safely
attached to the overhead rail. There had, alas, been cases where it
had not! I recall a case where a paratrooper was in this predicament
– next stop Mother Earth – when he managed to grab the line of
one of his fellows. The two of them landed – heavily – but got away
with it. I recall another occasion when a paratrooper whose chute
failed to open landed in a tree – and he lived to tell the tale.

Our other main duty was towing gliders. The glider pilots were Army personnel. They carried out a course of '*ab initio*' pilot training with the RAF. Then they trained with their RAF 'Tug' aircraft. It was the 'Tug' pilot's responsibility to position the glider relative to its landing-place. It was the glider pilot's responsibility to release his glider from the 'Tug' aircraft on a given signal. There were cases when release mechanism failed. One of the aircraft's crew then had to cut the towline – sometimes a hazardous operation.

I enjoyed squadron life. We liaised very closely with the Army, carrying out exercises simulating the type of operation we would eventually have to perform in Europe. In order to perfect the type of flying we would have to do, we carried out flights to various places in occupied Europe, dropping leaflets instead of paratroops. This was known as 'Nickelling'. One day I was summoned to HQ. 'We have a little job for you which should help your training. The French railways are very vital to Hitler. One in particular is the line from Paris to Bordeaux. As you know, the French railways are electric. This line depends on a transformer station near the River Loire. You are to attack it as if you are dropping paratroops – i.e. from 600 feet to 800 feet – but instead of paratroops you will drop bombs.'

We planned our route to avoid known places where there were strong AA defences. We flew at 12,000', in bright moonlight – we needed moonlight – until we saw the gleaming river. We descended to our dropping height of 800', turned to port and followed the river until we came to a bridge, then turned to port again – and the transformer station was about 400 metres away. We had been warned that the station would be defended, and – sure enough – I suddenly saw tracer coming towards our aircraft. For the uninitiated, a tracer bullet glows white when fired. One such bullet is therefore inserted at intervals in the belt of ammunition, thus giving the aimer an indication of the direction of his fire and enabling him to make corrections onto the target. I had never encountered it before, and recall feeling surprised at how slowly it seemed to be coming towards us – and (fortunately, in my case) missing! Our front gunner of course opened fire in reply; the bomb-aimer gave me final corrections: '...left... left...steady...Bombs away!' Then it was nose up, full throttle, turning to port as we climbed away. Looking back, I could see bombs from following aircraft bursting in the target area. We encountered a little 'flak' as we reached the Channel coast, but otherwise our return flight was uneventful. One of our aircraft failed to return. It was some time before we learned that it had been hit over the target area and had to crash-land. We heard later to our great joy, that the crew were uninjured; they had been well looked after by the French locals – at great risk to themselves – and eventually were smuggled over the frontier to Spain. Some of the other aircraft had sustained minor damage. Alas, I could fine none in mine!

No. 38 Wing was to expand rapidly into 38 Group and another Group, No. 46, was to be formed, as the airborne forces built up to take part in the invasion of Europe. But I was not to be with them.

One day Group Captain Harries, who had taken over the wing, summoned me to HQ. 'Donaldson, we want you to go to India to form a wing similar to this. It will be known as 177 Wing. Its immediate task will be to provide operational training for the 50th (Indian) Paratroop Brigade. They have completed their initial training, but have no aircraft for operations. Three squadrons of Dakota aircraft have been assigned to the wing – Nos. 62, 117, and 194. You will initially be based at Chaklala, Punjab. Make arrangements to leave as soon as possible. The post carries promotion to Group Captain.'

Line-up of Halifaxes and Hamilcar gliders for the invasion of Normandy.

My dear Sue and Sheila and I had found a small furnished house in Thruxton village. We had a WAAF billeted on us, and there were rather a lot of cockroaches, but we enjoyed it. Now we had to leave. We decided that the best thing would be for Sue and Sheila to go and live with her mother at Towcester. Mother was of course very pleased. It was a sad parting.

CHAPTER VIII

Back to India – Chindits

I flew out to India in stages: by Dakota to Gibraltar, then by another Dakota along the North African coast to Cairo. From there I was lucky enough to join a BOAC Flying Boat which took me to Karachi. A Dakota of No. 31 Squadron met me here, and completed the journey to Chaklala.

I found one of my squadrons, No. 62, already established at Chaklala. They had been supplying the aircraft for No. 3 Air Landing School, which had carried out the initial training of the 50th Brigade paratroops. Its chief instructor was posted while I was there, so I took over that unit too.

An important part of a paratrooper's training was carried out on the ground. A hangar was fitted with harnesses hanging from the roof. Simulated jumps were made from a gallery on a pulley arrangement, counter-weighted to approximate to the dropping speed in a 'live' jump. Due to winds, most landings would take place with 'drift' which could cause injury. One was therefore taught how to negotiate the parachute shrouds, not only to counteract drift as far as possible, but also to position oneself so that one could 'roll', with head and arms tucked in. I thought I'd better have a go, which I did, followed by a live jump. I obviously didn't do enough ground training, because I didn't roll properly, and gave my head a pretty severe bump on landing.

Squadrons Nos. 117 and 194 duly arrived and were stationed at airfields nearby. There came a period of getting to know each other and organizing our Wing HQ. As soon as possible I visited the brigade, then stationed at Campbelpore, Punjab, meeting their Commander, Brigadier Tim Hope-Thomson, and the COs of the Indian Army battalions forming the brigade. These included Gurkhas, Punjabis and Sikhs, and also signals and other supporting units.

There was a nice story about the Gurkhas, when they were being asked if they would like to jump into action from aeroplanes. After their officer had explained the procedure, there was a bit of a silence. Then the Gurkhas got into a discussion. Eventually a spokesman got up and said, 'Sahib, couldn't we jump from lower than six hundred feet – say fifty or a hundred?'

'No,' explained the officer. 'At a lower height the parachute would not open to let you down gently.'

Immediately they grinned happily and the spokesman said, 'OK, Sahib, we didn't understand about the parachutes!'

The brigade and ourselves immediately got down to operational training, organizing exercises simulating the types of action in which we would be engaged against the Japanese. A very good spirit developed between us outside our duties. The squadrons organized rugby and soccer matches against battalions of the brigade. One day I thought, 'Why not have a grand match between the Brigade (Army) and the Parachute Training School and 177 Wing (RAF), both teams to drop by parachute into the garrison ground attired in their football kit, ready for immediate kick-off?' We did it. We had to drop two at a time onto the ground, which was surrounded by a fairly substantial fence. As it was, I landed outside – probably again because of my lack of training. It was only my second jump! In the course of the game I sustained a knee injury which put me back a bit. News of this game apparently got back to HQ in Delhi, as a result of which we received an invitation from our Air Officer Commanding to fly

Combined soccer team from 50 (Indian) Parachute Brigade and No. 177 (Airborne) Wing, RAF, Chaklala, India.

With a group of Gurkhas, Chaklala, India, 1943.

our team down and 'drop' them into the Delhi stadium prior to playing a match against a representative team from the various HQs in Delhi. My knee was still bad, so I had to be content with flying the aeroplane. The occasion was a great success.

Before long I received a call from HQ Delhi. I was to go with the Brigadier to discuss a possible joint operation. There was a lot of 'top brass' at the meeting. They questioned us on our state of readiness. We replied that we considered we were ready. They described the latest developments on the South-East Asia front, where the Japanese were advancing steadily. Among the important places they had captured was the port of Akyab. It had been decided that we should make a determined effort to recapture Akyab. It was considered that our Airborne Force could be used to advantage in this operation. We should prepare for an early move over to the Bengal-Burma front. We returned to Chaklala excited at the prospect of at last going into action, and made preparations for an early move. Alas, for reasons which we never knew, the proposed operation was cancelled.

We had hardly got over our disappointment when we received further orders from Delhi. The position of Fourteenth Army was threatened by possible cutting of communications and supply routes. An operation was planned involving the Air Transport of a special force to strategic points in parts of the territory largely infiltrated by the Japanese. We were to leave immediately for various aerodromes in northern Bengal – Agartala, Hailakandi, Sylhet. On arrival I was ordered to report to the controlling HQ, 3rd Tactical Air Force, where I was briefed.

The operation of this special force was largely the brainchild of its commander, Major General Orde Wingate. His force had already carried out on foot a penetration of this part of Burma, during which he had noted a number of clearings in the jungle which made up most of the territory. The plan was to make these clearings into airstrips which would enable Dakotas to fly in, carrying his special force. Subsequently this force became know as the 'Chindits', named after the chinthe, the legendary beast of Burma. The flying side of the operation was to be a joint one with the US Air Force.

It is worth digressing here to describe briefly the Americans' role. Wingate's ideas had caught the imagination of the Americans, including General Eisenhower. They also had backing from Winston Churchill, himself a character who admired initiative and originality. Thus when Wingate boldly asked the American Chiefs of Staff for help, on President Roosevelt's instructions they assigned, to Wingate personally, a composite force called No. 1 Air Commando Group, under the command of Colonel Philip Cochrane, a legendary figure. It was, in effect, a miniature air force, comprising units of fighters (P51 Mustangs), bombers (B25 Mitchells), transport (C47 Dakotas) and Norsemen Helicopters (YR4), Gliders (Wako C6), and Light Communications Aircraft (Vultee L1 and L5). Wingate was able to call on Cochrane for action in any sphere – bombers or fighters to straf Japanese positions or aircraft on the ground. and to take him from A to B. Among these varied roles, possibly the most important was that played by the small 3-4 seater L5 aircraft in evacuating casualties. This aircraft could land and take off in some 200-300 yards. In the event of a serious casualty a clearing would be found as near as possible and a call made. The L5 would take the casualty to the nearest airstrip, where it would be transferred to a Dakota and flown to the nearest place where there were appropriate medical facilities. The effect of this service on morale was enormous.

Briefly the object of Operation THURSDAY, as it was named, was to cut Japanese communications between the south and their forces opposing General Stillwell and his Chinese Forces in the north, so that the important task of building the Ledo Road to

China would not be interrupted. If Wingate's 3rd Indian Division, the 'Chindits', could be flown into airstrips strategically placed, time would be saved, troops would be fitter, and surprise would be more likely.

Possible sites for strips had been noted during the first penetration. Strips were to be prepared by engineers flown in by gliders of Colonel Cochrane's Air Commando. As soon as the strips were ready, the Dakotas/C47s of Troop Carrier Command would fly in the Divisional troops and equipment.

The code names of the first two strips were 'Broadway' and 'Piccadilly'. On D-Day, 5 March 1944, thirty C47 aircraft were lined up, each with two Wako gliders in tandem behind them, loaded with airfield engineers, troops and equipment, including miniature bulldozers. Just before Zero Hour, a last-minute photo-recce revealed that the area destined for 'Piccadilly' had been obstructed with large tree trunks. Despite the possibility that this might indicate that the Japanese had knowledge of the operation, and might also be waiting for us at 'Broadway', Generals Slim and Wingate decided to continue, but switching the 'Piccadilly' contingent to 'Broadway'.

As darkness fell at Hailakandi, the first C47 took off with its two gliders, followed at five-minute intervals by successive aircraft. But trouble developed with the dual tow. Seventeen gliders broke away or had to release to crash-land, nine in hostile territory. The aircraft towing eleven gliders were recalled. But thirty-two gliders crash-landed at 'Broadway'. Such were the landings – some gliders having no option but to crash into one another – that thirty-three men were killed and thirty seriously injured. The miniature bulldozers were lost.

A few hours later the codeword 'Soya Link' was received, meaning 'Mission failed'. (For the uninitiated, a soya link was a horrible synthetic sausage.) We sent an aircraft over to drop more

74

equipment. Then we received another signal which spelt out 'Pork Sausage', meaning 'OK'. Those who had survived, having piled up their dead and succoured the injured, got to work often with their bare hands, and completed a magnificent job. On the night of D+1, sixty-two successful Dakota sorties were flown into Broadway, transporting the troops, mules and equipment of 77 Brigade, commanded by Brigadier Mike Calvert, together with part of 111 Brigade.

On D+2, a second strip was prepared, code-named 'Chowringhee', and twenty-four aircraft were airborne for it when a message was received, 'Strip only 2,700 feet long'. Allowing for their full load and trees at each end this was a bit short, and a recall message was sent to all aircraft. However seven aircraft 'did a Nelson' and landed – fortunately without damage.

By D+5, the fly-in of 111 Brigade, commanded by Brigadier Joe Lentaigne, and certain other units, was completed.

For their role of harassing the enemy, the Chindits split up into columns, each brigade comprising eight columns of 400 troops, plus mules and equipment. Two hours after the last column of 111 Brigade had left 'Chowringhee', the strip was bombed and straffed by Japanese aircraft.

In addition to Hailakandi, our aircraft operated from strips at Lalaghat and Tullihall. The latter was a long strip, but very dusty, such that an aircraft taking off raised a cloud of dust which delayed the next aircraft. There being little wind, we overcame this by positioning aircraft at the midway point, and taking off alternately in opposite directions.

The transport of mules is worth a mention. These ubiquitous if temperamental creatures had long been essential to military operations in this kind of territory. Bamboo stalls for five mules were constructed in each aircraft, and a straw-covered ramp led up to the door. The muleteer would coax the mule up the ramp and, hopefully, into the stall. I was taking a load of mules into 'Chowringhee' on D+2, when one refused to go up the ramp. Wingate was striding up and down in his somewhat impatient manner and immediately came over. Taking the bridle from the muleteer, he proceeded to 'walk' the mule round the aircraft, 'showing' it the various features. I understand that this is the classic psychology applied by the equine fraternity – showing a horse a jump before hurtling it at the obstruction. I would like to report that it worked. But alas the mule took even greater exception – much to Wingate's discomfort! Eventually it was manoeuvred in by six men and a rope. Some mules became very excitable in flight, and a number had to be shot before they kicked the side out of the aircraft. Nevertheless, they were valued so much that when their tasks were done, we flew most of them back out.

On this same night, Air Marshal Sir John Baldwin came along and said, 'Have you got room for me, Donaldson?' Of course we made room – at the expense of my Canadian navigator, Olaf Meyer, who then had to navigate standing up. Then Colonel Alison, who did a magnificent job sorting out the final debacle at 'Broadway' on D-Day appeared and said, 'OK if I join you fellers?' He squatted on the floor and promptly went to sleep – within kicking distance of a mule! I was very touched by this gesture, which so naturally epitomized the solidarity and confidence which had developed between our American allies and ourselves.

The jungle strips were shaped like the traditional thermometer, the bulb being the unloading area, able to accommodate five or six aircraft. After unloading, aircraft could only take off in the opposite direction. We seem to have ignored winds. Flying control was exercised by 'Red' and 'Green' Aldis lamp. By arrangement. we would split the night period between the American squadrons working from their airfields, and the wing aircraft coming from ours. On one occasion we had the period up to 23.30. I was the last aircraft off 'Broadway' at about 23.30. We were just starting our climb – at about 200' – when Olaf shouted, 'Aircraft dead ahead, Sir!' It passed directly over us with a dull roar, and our aircraft shook. In a 'sitrep' dated 9 March, Brigadier Tulloch, Wingate's Chief of Staff wrote of Troop Carrier Command's part, that it was '... a marvel of efficiency and smooth working'. I had one reservation!

During that first six-day phase, aircraft of the wing transported into these jungle strips 4,521 troops, 264 mules, 44 ponies, and 62,936lb of stores and equipment, and thereafter continued with supply-droppings.

By D+11, the Chindits had established a stronghold which they christened 'White City', cutting the Japanese road and rail communications to the north. Thus they had carried out their first objective.

On 25 March, in the second phase, 14 Brigade and 3rd African Brigade were flown into another strip, code-named 'Aberdeen'. But on this same day we were shocked to hear that General Orde Wingate, the inspiration of this and its earlier unique operations, had been killed in an air accident. Brigadier Joe Lentaigne, the commander of 111 Brigade, was immediately appointed his successor.

There followed the air supply of the Chindits as they moved through Japanese-held territory, harassing the enemy. A further stronghold was established, code-named 'Blackpool', where reinforcements could be landed and casualties evacuated. 'Blackpool' was attacked fiercely by the enemy. Anti-aircraft guns and medium artillery were brought up. Use of the strip became impossible, so supply had to be by dropping. On 25 May, the Chindits, carrying their wounded, broke out of 'Blackpool', evading the Japanese, to continue their task.

Meanwhile 14th Army, with its HQ in the Imphal valley, was

HQ 3 Ind Division
7 Ap '44

My dear Donaldson,

Congratulations on your D.F.C. which is small reward for the first class work you have put in during the last few weeks. Without your all out co-operation and effort this show would never have touched down, and I'm certain that it will be you and your aircrews and ground staff that will enable us to convert the try we have gained.

Few land men realize the strain of continuous night flying off and on to six different strips and over mountains and through the muck that you and your people have had to cope with. The two nights I spent with you trying to get to ABERDEEN taught me a lot, as also the three or four nights in early March when I was on the strip or being flown in. Above all this you had all the planning to carry out; and that I know was a big job well done.

Again all congratulations and the best of luck in the future.

Yours sincerely,

[signature] Lentaigne

"Joe" Lentaigne took over command of the "Chindits" after the tragic death of Wingate.

Letter from Brigadier Joe Lentaigne who took over command of the Chindits after Wingate's death.

78

```
                              HQ 23 Br Inf Bde.
                              c/o No.11 A.B.P.O.
                              DO/1.
                              5 Aug 44.
                              ------------------
```

Dear *Donaldson*,

        I cannot let the operational season close for us
without writing to express the very great admiration which
I and all under my command, who have depended on you for
their daily bread during the past three months, have. for
the way in which your crews and aircraft have overcome the
many obstacles in their way and never failed us.

        Your pilots have flown with a truly magnificent
disregard of danger and have repeatedly managed by sheer
determination to deliver the goods to us in the most
appalling conditions of weather and over some of the most
treacherous country in the world.

        All ranks in my Brigade would, I am sure, welcome
the opportunity of telling you and your people personally
how much they have appreciated this unfailing co-operation,
and of mourning with you the loss of those who unfortunately
lost their lives in keeping us supplied.

        From my Rear Component at your Base I have heard
nothing but praise for your willingness to help in every
way possible, and I hope that the version of your Service
motto - "Per Ardua ex Cumulo-nimbus" - which I coined in
the midst of the battle, may remain with you as a symbol
of a very happy and successful combined operation, and as
promise of re-union in the future.

        Yours *[signature]*

                              Perowne

To:-   Group Captain G.F.K. DONALDSON D.F.C., A.F.C.
       HQ 177 Wing R.A.F. (AGARTALA)
       c/o No. 12 A.B.P.O.

From:- Brigadier L.E.C.M. PEROWNE,
       Commanding 23 British Infantry Brigade.

Letter of appreciation from Brigadier L.E.C.M. Perowne,
Commander of 23 Infantry Brigade.

being besieged. We were ordered to airfields in the Chittagong area to pick up 5th (Indian) Division and transport them to Imphal to reinforce it. On completion of this, our main task was to fly supplies of food and ammunition into the valley, which was by this time cut off from ground communications by the Japanese. The valley was surrounded by mountains 6,000-7,000 feet high. When these were covered in cloud it was not possible to operate safely, as we had no radio-navigational assistance. We therefore used to 'stockpile' at an airfield called Kumbirgram just the other side of the mountain range, and work an intensive shuttle when the valley was clear of clouds.

Intensive air supply continued; the Battle of Imphal was won and the Japanese were on the retreat southwards. We maintained our supply role as they pushed the Japs past Mandalay and on to Rangoon.

Before the end of this phase I was posted to 229 Group at Delhi as Group Commander Operations – but I wasn't there long. I was ordered to form another wing, No. 118, at Rangoon. The role of this wing was to assist in Operation ZIPPER – the invasion of Malaya from the sea. ZIPPER never started. We heard that an atom bomb had been dropped on Hiroshima and that the Japanese had capitulated.

CHAPTER IX

Aftermath: Singapore, SE Asia, Australia, Japan

Since the fall of Singapore we had heard heart-rending accounts of the treatment of the prisoners taken there being made to construct a railway from Siam to Burma. We loaded our aircraft with food and 'goodies', and dropped them at the POW camps along the line of the railway, then flew on to Bangkok (Don Muang) to which the ex-prisoners were slowly proceeding, loaded as many as possible into each aircraft, and flew them back to Rangoon. I shall never forget the sight of these men waving to us as we flew over dropping supplies. Even from our height of some 600-800 feet, we could see that they were thin and emaciated. But still they had managed to make crude Union Jacks on the ground. It took several days for us to complete this phase. The ex-prisoners were taken to hospitals and hotels in Rangoon, where they were looked after royally. I flew Lady Mountbatten on several of these trips.

Our next orders were to proceed to Singapore. We flew across the Bay of Bengal, to Penang, a very pleasant island, thence on to Singapore, where we landed at Kallang. Our task was again to bring back ex-prisoners of war – mostly Dutch – from Sumatra and Java. We had been issued with dollars for any expenses we might incur, but on arriving at Batavia we found to our surprise that the locals would not exchange them. They had become used to the currency which the Japanese had printed. At our hotel this was soon rectified; the overjoyed proprietor issued us with wads of the Japanese-made money, and this was readily accepted! But the 'penny soon dropped' – and the locals eagerly sought our dollars. In a few days we had evacuated all the POWs back to Singapore, where preparations had been made to give them the best attention and care possible.

While these operations had been going on, our administrative staff had been trying to find HQ and living accommodation for various units. Initially we had stayed at the famous Raffles Hotel, which the Japanese had preserved very well. My Senior Administrative Officer eventually found and took over some rather fine bungalows belonging to the university. But this too was to prove a short-term arrangement. One morning I had to visit ACSEA HQ. Quite casually I met a senior officer, Air Vice-Marshal Bouchier, whom I knew slightly.

'Hello, Donaldson, what are you doing here?'

I told him, whereupon he said, 'How would you like to come to Japan with me? I'm forming the British Commonwealth Air Forces, part of the occupying force.'

'Thank you, Sir; I'd be honoured.'

'The Forces will comprise two squadrons of RAF Spitfires, two squadrons of Royal Indian Air Force Spitfires, two squadrons of Royal Australian Air Force Mustangs, and two squadrons of Royal New Zealand Air Force Corsairs. We will be based at Iwa Kuni, which is about twenty miles from Hiroshima. My deputy will be an Australian Air Force Air Commodore; you will be Group Captain Operations, responsible for the co-ordination of the work of the squadrons. You will have to go to Melbourne, Australia to meet the Air Commodore and co-ordinate that side of it. The Indian and RAF squadrons will assemble at Seletar (Singapore) and be taken to Japan in the aircraft-carrier HMS *Vengeance.*'

When he had finished this briefing I said, 'This is obviously a Tactical Air Force, Sir, and I have no experience in this field.'

'Never mind,' he replied, 'I went through the Battle of Britain, and can supply that element.'

Indeed he had been – as a distinguished station commander in 11 Group covering London and South-East England. AVM Bouchier, affectionately known as 'Boy Bouchier', was a very shrewd and able officer, his efficiency cloaked by a nice sense of humour. We seemed to get on well together.

The two RAF squadrons, Nos. 11 and 17, were based at Kuala Lumpur, so I flew up to make their acquaintance. No. 17 was commanded by the famous 'Ginger' Lacey, who holds the distinction of shooting down the most enemy aircraft as a sergeant pilot in the Battle of Britain. Not being conversant with the Spitfire, I did a few circuits and landings. Ginger warned me that the mark of Spitfire – 19, I think – had a strong tendency to swing to the right on take-off. I found it so, but after an initial over-correction it was OK. I found it also swung after landing.

Then my job was to get to Australia. In a few days, 'Movements' called me to say that a Dakota was calling en route for Australia, so I packed a bag and boarded.

On board I was very pleased to meet an old AOC of mine, Air Marshal Sir Alec Coryton, who was going to take up a senior post in Australia. I recall that the pilot was Group Captain Grindell, commonly called 'Grinders'. He was an Australian, and very keen on horse-racing. The Melbourne Cup, the equivalent of our Derby, was due to take place in Melbourne just before Christmas, and he estimated he might make it if he 'opened the taps' and didn't waste any time. We stopped at various aerodromes en route and eventually landed at Darwin in North Australia, thence via Cloncurry to Mascot, Sydney and on to Essendon, Melbourne. I remember Cloncurry, where we night-stopped, as a near replica of the one-horse towns depicted in Western films – horses tethered to rails outside shops and bars, men wearing typical wide-brimmed hats.

Melbourne is a lovely city, and we made it in time for the 'Cup' which was the next day. I had the privilege of going with Sir Alec, and I recall that we were very lucky, backing three winners!

After our conference with the representatives of the RAAF and RNZAF contingents, we returned to Singapore. Our return flight was notable for one unusual occurrence. We were about to take off after refuelling from an airfield in Labuan. Just before the pilot opened the throttles, there was quite a violent rocking of the aircraft. He thought it was an airman who was at the wing-tips, as rocking the wing was a method of drawing the pilot's attention to something. But the airman just opened his hands, indicating that he had no part in it, whereupon the pilot opened up and we took off. Days later we heard that there had been quite a serious earthquake in that region.

Back at Singapore Nos. 11 and 17 Squadrons were assembled at Seletar in the north of the island, awaiting the arrival of the Aircraft Carrier HMS *Vengeance*. It had first of all to pick up the two Royal Indian Air Force Squadrons at Cochin on the west coast of India. When it arrived, the aircraft were hoisted on board. There was no practical way of flying them on, nor had we the experience to do so; nevertheless we still hoped to 'fly off' when we reached Japan.

On the first evening aboard, I received a summons to the Captain's cabin.

'Donaldson, you will appreciate that you and your officers just about double our normal complement of officers in the wardroom, so we'll have to ration you on the drinks. I've worked out that we can manage four bottles of beer and two large whiskies per day. Gins are OK.'

HMS Vengeance.

After the privations of Bengal-Burma when we were lucky if we got beer once a week and perhaps a bottle of gin (Indian) or whisky (Australian) at uncertain intervals I did not complain.

The voyage took ten days, spent interestingly enough getting to know each other and our naval colleagues. The flight deck provided us with ample open space for exercise. Sometimes we played deck hockey, sometimes just walked up and down. 'Ginger' Lacey was a frequent companion on our walks. We still hoped we might do our first deck take-off on arrival, and he would look at the structure of the funnel and bridge on the starboard side and say, 'Remember, Sir, the Spitfire 19 swings like a b— to the right on take-off.'

Our base was Iwa Kuni, on the inland sea. Any idea of flying the Spitfires off was soon abandoned for various reasons. However, we had a small L5, rather like an Auster, for communications purposes. I flew this off without much difficulty or skill. So I *can* say I have done a deck take-off.

As soon as we had settled in we had, of course, to visit Hiroshima. It was really an awesome picture of complete devastation. I shall never forget seeing, on a fragment of white wall, the dark silhouette of a human figure.

We found the Japanese friendly enough. They supplied us with servants in the Mess, including batmen. But they were batwomen! They took it as part of their duties to run one's bath, and scrub one's back, with complete indifference.

One day I took our L5 to visit 17 Squadron, based at an airfield about sixty miles away – I forget its name. The weather was clear when I took off, and the forecast good. But as I approached I ran into what seemed to be thick cloud. Then I realized it was smoke! When I emerged a few minutes later I looked around and saw a conical mountain on a small island with smoke and flames pouring out of its peak. It was an amazing sight, as the molten lava poured down the sides and into the sea where it raised great clouds of steam. When I landed I heard that the volcano had only started to erupt about twenty-four hours earlier.

Before long demobilization grew apace, and I received orders to return to England. There was no ship or recognized route, so I managed to board a Dakota bound for Melbourne. By this time rations were easing and the NAAFI was quite well stocked with drink. I managed to purchase a case of Scotch which was put on board. Our route was virtually that taken by our American Allies in their advance on Japan, and we night-stopped at several of their aerodromes. They were most hospitable and friendly. At one place we were having a drink and chatting, when one of the Americans took out a pen to sign the drink chit. I noticed it was a rather fat black instrument. It wasn't a fountain pen or a pencil, but the writing was distinctive. It was the first ball-point pen I had ever seen! I was fascinated, asked if I could try it – and offered him a bottle of Scotch for it. He accepted with alacrity. It stopped writing about a week later!

Our first landfall in Australia was Darwin. From there we flew to Fremantle. I remember this town as a replica of the mid-west towns of cowboy stories; horses tethered to posts in unmade roads, wooden shanty-type buildings. But the hospitality was wonderful. Then Sydney, the capital, a beautiful city, with its long bridge over the harbour. There was an air of excitement when we arrived; I soon found out why – there was an England Rugby League side touring, and it was the day of the first Test Match!

The match was to be played on the famous cricket ground, venue of many notable Test Matches. I joined the crowd; I had no ticket, of course, but hoped to get in. I need not have worried, I was in RAF uniform, and as soon as the officials manning the gate spotted me they shouted, 'Come on, Cobber'. I was hustled up to the centre stand. The occupants of one of the best rows were told to 'shove up and make a place for a Cobber' – and I was in. Everyone had a bottle, or cans of beer, which they passed to me freely. My recollections of getting back to my hotel are hazy...

I stayed in Sydney for two days, and then on to Melbourne. Here I met Rob Davy, the CO of 62 Squadron of the Wing. He introduced my into the United Services Club, where I was able to stay; it was most friendly. One day I was introduced to the Managing Director of Victoria and Inter-State Airlines. He said, 'What are you doing when you get home?'

'I don't know,' I replied.

'Well,' he said, 'how would you like to work with us?'

'I should like it very much,' I replied, 'but I must of course go home and discuss it with my wife.'

'Of course,' he said; 'let me know.'

After a day or so in Melbourne, I was notified that a passage had been booked for me on the SS *Caernarvon Castle*, sailing from Fremantle, near Perth. So I flew to Perth and duly joined the vessel. It was under contract as a troopship, which meant that it was 'dry', so my case of Scotch was much appreciated – but, alas, it didn't last long. I had a very comfortable cabin, and soon got to know some of my fellow passengers, mostly on repatriation like myself. Our route was across the sea until we reached Suez, went through the Canal, down the Mediterranean, round Gibraltar and home. Sue met me at Waterloo – and did we have a big hug! Then it was back to Sue's home at Towcester, where Sheila, now three years old, was waiting.

We had let our cottage in Perth, but now we were able to reclaim it and moved up there to enjoy a few weeks together and to contemplate our future. We decided not to go to Australia, so I duly wrote to Victoria and Inter-State Airlines thanking them for their offer but declining.

I had once met Henri Plessman, sometimes known as 'the Father of KLM', the Dutch airline. I wrote to him asking him if there were any pilot vacancies, and received the reply, 'Come over to Amsterdam and see us.' Armed with the letter, I presented myself at the KLM desk at Prestwick Airport, and was put on the next plane to Amsterdam. I was offered a job subject to completion of a satisfactory period of training, and returned home to discuss it with Sue. Then I heard that the newly-formed Ministry of Civil Aviation was looking for people with flying experience to form an Operational Directorate. I applied, went up to London for interview, and was offered a job as Operations Officer Grade I. I was now approaching thirty-nine years of age –

perhaps a little old to start an airline career, so I accepted the offer – and became a civil servant. The job was at the HQ in London; we had to move from our nice cottage in Scone to a flat which we found in Winchmore Hill.

CHAPTER X

Civil Servant

Civil Aviation was still relatively undeveloped and our main job was to formulate regulations for air safety, air traffic control, aerodromes, aircrew licensing and so on. It involved liaison and meetings with bodies like the Royal Aero Club, British Airline Pilots' Association, Charter Pilots' Association, airlines and aerodrome authorities, some of which were under government auspices, some municipal and some private.

The realization that aviation was breaking down the frontiers of the old world made it obvious that co-operation between all countries was essential. The United Nations Organisation had by this time come into being. It comprised agencies covering the main activities which required international co-operation and agreement, e.g. food and agriculture (FAO), health (WHO) etc. and aviation. The aviation agency was named the international Civil Aviation Organisation (ICAO) but in the early days it was prefixed with 'Provisional' (PICAO). Its role was broadly to agree safety standards and uniform procedures as between all members, in the international operation of civil aircraft.

I was nominated a member of the British Delegation to the second PICAO Conference in Montreal. The delegation comprised representatives from all facets of aviation – airline pilots (BALPA) and operators (GAPAN), BEA, BOAC, flying clubs, charter pilots, the Royal Aero Club, gliding, the Glider Pilots' Association etc. The Department (then the Ministry of Civil

90

Aviation) held meetings at which the UK brief was hammered out. We had as our leader the legendary Sir Frederick Handley Page.

Our travel arrangements were left to our individual choice. The options were by sea or by air. I made enquiries and found that there was a convenient sailing by the *Queen Mary* to New York, thence a flight to Montreal. With some feeling of disloyalty I opted to go by sea. I duly reported to Southampton, dumped my suitcases in my cabin (first class!) and repaired to the forward bar. There, one after another, assembled the whole of the delegation – all had had similar thoughts!

The voyage was a memorable experience, particularly from the food point of view. We were still suffering a degree of rationing at home, but not on board! At dinner I was sitting next to Sir Frederick who was something of a gourmet. I recall that on the first night we were tucking in to a liberal portion of smoked salmon, when the wine-waiter asked me what I would like to drink. 'A pint of beer,' I ordered. 'You can't have that George!' Sir Frederick said '– it's sacrilege.' He thereupon ordered a bottle of vintage white wine – and thereafter gave me lessons in what to drink with one's food.

The conference proved most interesting and constructive. It was conducted in English, but with instantaneous translation into the language of members who did not speak English – who were connected via headphones to an interpreter. I made the acquaintance of many of the other delegates, of the seventy or so member states. Nor was the importance of the social side neglected. There was a cocktail party every evening – sometimes two in a night!

The following year there was a further conference, again in Montreal. I was invited to be leader of the UK delegation – an honour which I greatly appreciated. One of my abiding memories of this conference was the mutual friendship and co-operation which quickly developed between the leader of the US delegation, Bill Flener, and myself. It soon reached the stage where we would go out to dinner – after the usual cocktail party – and discuss the next day's programme. Almost invariably we agreed the line we should take and carried it through at the next day's session. There was only one occasion when Bill said to me, 'George, I'm sorry I can't agree with your line on this one.' The subject was 'Take-off and Landing'. Experience throughout the world had demonstrated that, although individual Aircraft Performance Regulations required that an aircraft should not use a runway unless it met the minimum length specified for the type, aircraft sometimes overran the runway – and suffered damage and casualties. The UK brief was that 'Beyond the end of every runway there should be an over-run area free from obstructions and capable of supporting an aircraft without its incurring serious damage if it overran.' The reason Bill could not support this recommendation was that the US Government was pledged by law to meet the costs of all mandatory aerodrome developments. With their great and ever-increasing numbers of aerodromes, this would have been a heavy financial commitment for which he did not have a mandate. (Nevertheless this safety measure was adopted at the next conference!)

Measures which were eventually agreed were categorized as either 'Standards' or 'Recommended Practices'. 'Standards' were to be incorporated in the legislation of all member states as mandatory; 'Recommended Practices', as the name implies. were optional.

Looking back, it was a privilege to have been in on the development and framing of international standards governing civil aviation, and to see how they worked and were further developed during the succeeding years – and to recall that some of them had been worked out with Bill over a pint in a Montreal pub!

When I got back from the conference, I found I was not to be long in HQ. The Ministry had decided to decentralize into divisions – geographically. These were to be Scottish, Northern, South-Western and London. Each Division would deal with day-to-day administration of stations in its territory. I was posted to Scottish Division centred at Prestwick, as the Chief Operations Officer and Deputy Divisional Controller. Our territory covered aerodromes and radio stations in Shetland (Sumburgh), Orkney (Kirkwall), the Hebrides (Stornoway), Inverness, Aberdeen, Perth, Edinburgh, Glasgow, Prestwick etc.

Our HQ were located in a large mansion on the perimeter of Prestwick Airport, situated in grounds of rhododendrons etc. Living quarters had been built in the grounds, one of which was assigned to us. It was a detached building, set in trees. My 'journey to the office' was a five-minute walk through rhododendrons and across a lawn. My office was a former drawing room of the house.

Our family life was pleasant here. Sue made a number of friends and indulged herself at her favourite sport of tennis, But she had to slow down because of an impending important and happy event. Alastair was born on 26 July 1948. Sheila was by this time six years old, and was attending the local school in Prestwick. As was natural, she soon spoke with a distinct Scots accent. I recall her coming home one day and reciting a poem she had learnt. It went:

> Aw dandelion, yellow as gold
> What do you doo all day?
> Ah sit around in the tall green grass
> And watch the birdies play.

My job was also pleasant. We all got on well together socially and in our work. The latter entailed visiting all the units in the division. Normally this would have involved a lot of journey time, but HQ decided to keep a small detachment at Prestwick – usually

Sue outside our cottage in the grounds of Adamton House, Prestwick.

one aircraft, an Anson of the flying unit from Stansted. Its main job was to calibrate all the Telecom stations, and this took up most of its time. But I was sometimes able to borrow it, or to co-ordinate my visits with the calibration work.

I have always paid tribute to the Civil Aviation Department's policy of making sure that their technical officers, on whom they depend for advice on the framing of legislation ensuring air safety, efficiency licensing etc. were given the facilities to keep in touch and up-to-date with the rapid developments in civil aviation. Thus they maintained a flying unit at Stansted (of which I had the privilege of being commander, for a time, on secondment.) The unit contained twin-engined aircraft – the DH Dove and the Avro Anson; also single-engined aircraft – the Auster and, for a time, a Tiger Moth. The unit's two main functions were:

94

(i) to ensure that the standard of flying for pilots' licences and ratings was of a high and uniform standard. Initially all flying tests had to be carried out by examiners of the unit, but with the expansion of aviation this became too great a task, and examining was delegated to schools and operators. Delegated examiners were given a short course of training by the unit and were subsequently subject to spot checks – as were their 'products'. This was conveniently done by a Ministry examiner flying with the delegated examiner while he carried out a test.

(ii) the checking of all radio-navigational and landing facilities. Additionally the aircraft enabled Ministry staff to maintain their flying competence and licences, and also provided communications for staff visiting aerodromes.

Thus one could 'keep one's hand in' and maintain one's pilot's licence. Before the War (1939-45) there had been two pilot's licences, the 'A' for private flying only, and the 'B' for professional flying. After the War the licences were revised. The 'A' licence became the 'Private Pilot's Licence' (PPL). For commercial/ professional flying, three licences were introduced, the Commercial Licence, the Senior Commercial Licence, and the Airline Transport Licence. To qualify for each licence one had to pass an examination, and one had to have the prescribed number of flying hours as pilot in command. My 'B' licence was converted to an Airline Transport Pilot's Licence without examination – for which I was very thankful!

Other courses available were for gliding and for flying helicopters. I applied for a glider course. It was run at Perranporth, an aerodrome right on the edge of the cliffs in Cornwall. I took Sue with me, and we had a delightful three weeks 'holiday' on this lovely stretch of the South Coast. At the end of the course I obtained my Glider Pilot's Certificate and a little knowledge of how the gliding community operated – which was why I had gone on the course!

The following summer I was granted a 'refresher' course, so back Sue and I went one lovely June day, to Perranporth. Now there are two ways of launching a glider. It can be towed off by another aircraft and taken up to any height required. If the glider hopes to do a cross-country flight it would probably be towed up to 3,000 feet or so, and released. For 'circuits and landings' – or such local flying as the availability of 'lift' invites – the glider is towed behind a car, which belts along the runway achieving a speed of around 40-50 m.p.h. The glider becomes airborne in a very short distance and thereafter the pilot moves the 'stick' back and climbs at just above his stalling speed. The principle involved is identical with that of a boy running as fast as he can, towing a kite. When the car reaches the end of the runway, the glider is probably at around 600' and of course has to release the tow-rope. This height is sufficient for the pilot to make a circuit and land on the runway he has just left. On a good gliding day with plenty of cumulus cloud about, he may find a 'thermal' and find his glider gaining height. This might encourage him to venture further afield.

I hadn't been up in a glider since my visit a year before, and fully expected that I would be given a check-flight with an instructor. However the Chief Flying Instructor said, 'George, I don't think you need a check; off you go in G-ABCD.'

The duty runway was '26'. It headed straight towards the sea, finishing virtually on the edge of a cliff about 200' high, with the sea beneath. After the usual signals, the tow-car accelerated, the tow-rope tightened, and I concentrated on keeping straight. We became airborne and I watched for my speed to build up. Suddenly I realized that the car was at the end of the runway. I had to release the tow-rope. The cliff edge was beneath me with the sea beyond. I was only at 200' instead of the 600' that I should have been! The classic doctrine – which I had expounded to hundreds of pupils and embryo flying instructors – and as laid down by the Central Flying School in respect of 'taking off' is, 'If your engine fails immediately after take-off, put your nose down, achieve gliding speed, and try to land straight ahead... Do not attempt to turn back to the aerodrome' Here was a classic case – only instead of the engine having failed, I did not have an engine! If I followed the time-honoured injunction, I would have to land in the sea. If I turned back, I might hit the face of the cliff! On an instinct I turned back. I cleared the cliff-top by a few feet and landed, downwind, but safely on the runway.

The 'post-mortem' (fortunately in its less sinister connotation) soon explained my unforgivable error. Paradoxically it again has to do with the golden rules of flying instruction in fixed-wing aircraft. The procedure for take-off includes 'as soon as the aircraft becomes airborne, move the "stick" forward and fly level until climbing speed has been attained.' This is what I had done instinctively, forgetting that I was being towed in a glider. I had flown level waiting for my speed to build up. Of course it was unable to build up beyond the speed of the towing car! That took a bit of living down!

Probably the most valuable and interesting course available was a secondment to one of the corporations – BEA or BOAC as a pilot. To be eligible for this, one had to possess a valid professional pilot's licence. At the ripe old age of fifty-three I had not considered I was eligible. But one of my colleagues, who had been nominated, had to pull out for family reasons, and the director responsible said, 'George, would you like to go?' Of course, without hesitation, I said yes.

CHAPTER XI

Airline Pilot

I was nominated for British European Airways and attached to the Viscount flight. The first duty was to learn all about the aircraft. BEA ran a most efficient school near Heathrow, and I joined a course of some twenty-five pilots also destined for the Viscount. At the end of the course there was an examination under the auspices of the Air Registration Board. I passed. These were the early days of the flight simulator, and there followed training in all facets of flight in a most realistic atmosphere. While this was going on, I was fitted out with a uniform – with the two broad stripes of a First Officer. Then followed a course of flying training on the actual aircraft. It was carried out in Guernsey. After a very thorough flight test, day and night, I was passed, and my ALTP licence was endorsed as 'competent to fly the Viscount 701 as pilot-in-command'.

Initially I was based at Manchester and flew all the domestic routes which included Glasgow, Belfast, Dublin, Liverpool, Birmingham, London; also Paris and Dusseldorf. Some schedules necessitated getting up at 5 a.m. Pre-flight preparation was very thorough. It included visits to the Air Information Service (AIS) where there was up-to-date information on anything which might affect the flight – e.g. military activity, serviceability of radio facilities, danger areas, serviceability of aerodromes en route. Then a visit to the Meteorological Office, where one received a full briefing from the Meteorological Officer in charge. This of course included the winds likely to be encountered, which would affect one's flight time; cloud conditions, which could affect the flight level selected; storm areas, which it might be desirable to avoid etc. Checking the aircraft was also an important duty. As aircraft and instrumentation became more complicated, 'checklists' were instituted. Either the Captain or his First Officer would read out the various items, and the other would check them. Important manoeuvres such as the take-off virtually required both pilots. The one would man the flying controls, the other checking all the appropriate instruments, calling the speeds 'V1' – speed at which the aircraft has flying speed and the pilot 'rotates', i.e. lifts the aircraft off the runway; 'V2', the speed at which climb should commence. If, during the take-off, the co-pilot observes any irregularities or warning lights, he calls for the take-off to be abandoned. Minimum runway lengths are such that if the take-off is abandoned before the aircraft becomes airborne, there is room, using maximum brake and reverse thrust, to bring the aircraft to rest before reaching the end of the runway. Similar checklists are provided for approach and landing, emergencies, etc. I was allowed to do most of the flying and thoroughly enjoyed it. One day after a trip I was told that the Chief Pilot, Captain 'Bill' Bailey, wanted to see me. I had known Bill when he was Chief Pilot for the Scottish routes, and we got on well together. 'George,' he said, 'as you know this arrangement with the Ministry does not allow you to become a Captain, but I can promote you to Senior First Officer, and would like to do so.' I was pleased at the gesture, and had the thin gold stripe of the rank inserted between the two broad stripes. Alas, the period of my full-time attachment came to an end, and I returned to my desk at the Ministry.

In addition to my liaison with BEA, I had got to know one or two charter operators quite well in the course of routine checks of their instrument rating delegated authority. One of these operators based at Croydon had a contract to fly newspapers to major European cities – Amsterdam, Brussels, Paris and so on. 'Next-day's papers, hot from the press', would be rushed out to Croydon round about midnight, and loaded with great speed into the waiting aircraft, which would take off to arrive at their European destinations in the early hours. I asked them if, subject to the agreement of my Department, I could occasionally fly with them as a crew member. They said they would be very willing, as it would enable them to give one of their pilots a 'night off' on such occasions. My Department agreed. The best night was Friday, as I then had the weekend to catch up on my sleep. There were one or two days in the year when this liaison was particularly enjoyable. These were when there was a Rugby international – France v England or Scotland – on the Saturday.

My great friend of those days, alas no more, was one Dudley Smith. Dudley had commanded a battalion of the Essex Regiment in the epic landing on Anzio beaches in Italy in the War. Though severely wounded in the stomach, he carried on. He was awarded the DSO for great gallantry. After a long spell in hospital he was discharged with a permanently damaged liver. He was told, *inter alia*, that he must not touch alcohol. Dudley did not like the prospect of not being able to have a drink with his friends, so, not being short of a penny or two, he started going to various specialists with a view to getting this ruling overturned. Eventually one whom he consulted said, 'What were you in the habit of drinking, Colonel Smith?' 'Oh,' said Dudley, 'perhaps a couple of pints of beer at lunchtime – and two or three gins in the evening.' 'Well,' said the specialist, 'I don't think these would do you harm.' Dudley visited no more 'specialists'. He enjoyed life to the full and from my personal knowledge was none the worse for it.

Dudley and I had in common a love of Rugby, and used to go to most of the international matches. When England or Scotland

Dudley Smith and DH Comet.

were playing France in Paris, I would ask if I could fly on the newspaper flight on the night before – and could I bring a friend with me? The answer was always yes. So, having got what we called a 'pink ticket' from our dear wives, we would meet after work on the Friday, have a leisurely dinner, and hie ourselves out to Gatwick. Latterly the aircraft used was the ubiquitous Dakota. Its empty fuselage would be stacked up with great bales of the *Daily Telegraph, The Times,* the *Daily Mail* etc. It was always a source of interest how many English newspapers were read in Europe. Dudley would be squeezed in on the floor between the bales; I had a seat in the cockpit – and off we would go.

Arriving at Le Bourget in the small hours of Saturday morning, we would usually find a room at a hotel called the 'Oxford and Cambridge'. Dudley had been at Cambridge, and had adopted this hotel from earlier visits. After the match we had ample time to see Paris by night before getting ourselves out to Le Bourget to join the return flight of the aircraft which had brought in the Sunday papers! We would arrive back at Croydon at about four or five a.m. on the Sunday – and home to our ever-loving wives in time for an early breakfast – and bed!

CHAPTER XII

'Uncle George'

While I was in the Scottish Division in 1950, Prestwick Airport was on our doorstep. It was then very important to European as well as British airlines as a refuelling and staging post to North America. Sometimes Sue and I would go down in the evenings. One could meet and converse with aviation folk, and often meet old friends over a drink in the bar. One evening I got into conversation with a young fellow and asked him if he was a pilot. 'Yes,' he said. 'I'm flying a Taylorcraft up to Croydon – hoping to sell it.'

'How much are you asking for it?' I asked idly.

'£400,' he replied.

After another drink I said, 'I'll give you £350 for it.'

He accepted. We went over to the hangar, had a look at it, I wrote a cheque for £350 – and became the proud owner of G-AHUG – or, in the phonetic alphabet of the day, 'George Alpha How Uncle George'.

Sue and I were to have a lot of fun with 'Uncle George' in the succeeding years, but at the time I was thinking of how I could use it in my job of visiting Scottish aerodromes. As in many organizations, there was in the Ministry an arrangement whereby one could use one's own private car for duty journeys, and claim mileage allowance. The allowance in those days varied according to the horse power of one's car – e.g. for an Austin 7, sixpence a mile – up to tenpence a mile for a large car. I thought, 'Why shouldn't I make a case for using my aeroplane and claim allowance for it?' I did. At first it was pooh-poohed, but we had a very good Departmental Secretary, Sir Alfred Le Maitre, with whom I got on quite well. My case was apparently referred to him, for one day he phoned me and said, 'Donaldson, would you accept the top rate for cars i.e. tenpence per mile?' I made a quick calculation. My average cruising speed was about 80 miles per hour – at tenpence this meant 800d, or £3.6.8 per hour. My aeroplane cost £3 per hour to run! I accepted. This concession not only saved time, but I found it was appreciated by airport authorities, on the grounds that 'this chap has practical knowledge of the facilities being provided'.

Apart from using Uncle George on duty, I soon found that several of the staff were keen to learn to fly. As I still retained my instructor's rating, I said 'Why not?' There were the Divisional Telecommunications Officer, the Divisional Fire Officer, my ops staff, a KLM executive, and a business friend who had been a navigator in the RAF – I gave them all lessons in our spare time, and they all duly went solo and qualified for their Private Pilot's Licence.

On the sporting side, the Royal Aero Club organized a number of events for clubs and private owners of light aircraft. Among these were the National Air Races, the King's Cup, the Grosvenor Trophy and so on. I thought it would be good fun to enter for these and indeed it was. I see from my log books that I participated in the Grosvenor Trophy 1949 – in which I finished fifth – and again in 1950 – when I was sixth. I took part in the Daily Express Meeting, Shoreham in 1951, the Goodyear Trophy, Wolverhampton in 1952, the Ragosine Trophy in 1952...I did not go to great lengths to improve the performance of Uncle George – just a bit of polishing, and the removal of anything which one could legitimately remove. The aircraft's normal cruising speed was about 80 m.p.h. I see from my log books that in the Grosvenor Trophy in 1949 my average speed was 104.25 m.p.h. This is accounted for, of course, by the fact that one was at full throttle all the time!

In the Grosvenor Trophy of 1950 I remember being overtaken and passed by a similar type – Taylorcraft 'plus D'. I was a bit surprised. I was acquainted with the owner and remarked on it after the race, in which he had come third. He explained: 'I removed the ballast weights from my tail, old boy.' Strictly against regulations, of course, but he got away with it. And why not?

Another great pleasure we experienced with Uncle George was going on rallies abroad. These were usually sponsored by well-known businesses and were attended by private owners and club syndicates from the UK, France, Belgium, the Scandinavian countries etc. One I recall vividly was the Cognac Rally. It was sponsored, inevitably, by the various distillers of the brandy which bears its name. Sue loved these occasions. She would take the controls and flew quite well. But mainly she liked to map-read for me. Woe betide me if I was the wrong side of a railway line on our track! On this occasion our route was Sywel-Lympne-Le Touquet-Deauville-Angers-Cognac. The final leg – Angers-Cognac – was timed, and prizes were awarded on a handicap basis. On arrival we were marshalled onto the tarmac. Just as we were getting out, I glanced at the aircraft which had stopped alongside. Its registration was OO-FUN – and we were phonetically Gee – A hug! We all had a jolly good laugh and entered the hangar, where we were welcomed by our host and introduced to the hospitality of all the brandy-makers of Cognac, every one of whom had their own hospitality room! We stayed at the chateau of Jean Martel, sleeping in a truly old room, in a four-poster bed. The next day there was a competition. We were each given a map of the district and told that the insignia of every brandy-maker were displayed at various points in the countryside, which was largely wooded. We had to spot these and plot them on our map. Having done so, we were to place the map in a weighted bag, return to the airfield and dive-bomb a target in the centre. Sue was the bomb-aimer. She had to open the door and throw the bag out when I gave her the signal. She was a bit scared, but it all went according to plan. Our bag finished up about twenty-five yards from the centre – a result which won us a few bottles of cognac. We were given other bottles, including a collection of miniatures. After a very good weekend we returned, landing in the UK at Lympne where we had to clear customs. We declared our gifts of cognac, but alas were charged duty!

We also toured Europe. I recall a flight from Marseilles. Our route was up the Loire valley. I was flying at about 3,000 feet at my cruising speed of about 80 m.p.h. when I began to notice that we were hardly moving relative to the ground beneath us. We were heading into a Mistral! I then recalled from my study of meteorology that this wind blows strongly from the north in summer. It is caused by heated air over the Sahara rising, and thus drawing in cooler air from the north. I reduced height to about 1,000 feet, but was still making little headway, so I decided I would have to land. I saw from my map that we were not far from Montelimar, famed for nougat, so I aimed for it. The air was now very bumpy, and when we reached the airfield I noted that the windsock was horizontal. I estimated that the surface wind must be 50 m.p.h. probably gusting to 60. This was far above our safe landing speed. I circled the airfield at a fairly low height, and soon about four or five people emerged from the hangar and clubhouse. They came out onto the airfield and positioned themselves either side of my landing run – about a wing-span apart. I got the message, and motored in between them, closing my throttle and landing at the appropriate point. They grabbed my wings and tailplane, and escorted me in without damage. We stayed in Montelimar as their guests. They were lovely people. We managed to get away the next day with their good wishes and a few pounds of nougat!

Back at Prestwick, life continued as usual; Alastair was now two years old and a fine boy. We still lived in the 'quarter' in the grounds of Adamton House. Round about 5 o'clock, I would frequently hear a small voice crying, 'Daddy, it's time to come home.' There was Alastair, having trotted the couple of hundred yards from our house. Not many people can have had this experience. I did not keep him waiting!

In 1951 I was posted to Stansted as Commander of the Civil Aviation Flying Unit. Having some fifteen aircraft of various types under my command and the job being static, the need for 'Uncle George' diminished; however I continued to enter for events, often taking Sue with me. I had two hats at Stansted. In addition to the main job of the Flying Unit, the Powers said, 'You might as well be commandant of the airport.' In those days there was not a lot of traffic other than that of the Flying Unit. Freddie Laker was beginning to make his mark, but initially it was as a converter of RAF Lancaster aircraft to civil use. One of my 'perks' at Stansted was Renfrew's Farm as an official quarter. It had originally farmed the whole area. It was situated about 200 yards from the end of the main runway, well inside the safety limits as laid down today. If we had a party, some of our friends would arrive in their private aircraft, and park just outside the front door!

One day we met a couple of young fellows who expressed great keenness in learning to fly. I said, 'I have an aeroplane. I will teach you. If on completion of your courses you would like to buy my aeroplane, I am ready to sell it.' They thought this was an excellent idea, and agreed. Every weekend they used to come up from London with glamorous girlfriends. These were handed over to

Sue and 'Uncle George'.

Sue to look after and help with the tea, while we went off to fly. They were apt pupils and in a few months became qualified and passed their examinations for the PPL. True to the spirit of our bargain, they said,

'Right, now, how much do you want for Uncle George?'

'£400,'I said.

The bargain was sealed. With some sadness on both Sue's and my part, we waved goodbye to Uncle George.

But this was not the end of having fun flying light aircraft. The aircraft of the Flying Unit were maintained by a civilian contractor, Helliwells. One day the manager, with whom I got on quite well, said, 'George, we've acquired a small aeroplane. It's called a Globe Swift. We imported it from America hoping there might be a market for the type. If you would like to borrow it to fly in the National Air Races, you are very welcome.' I was intrigued, and accepted their kind offer.

The Swift – AHUU, 'Uncle Uncle' was a low-winged monoplane considerably faster than the Auster. I flew it in the

The Globe 'Swift'.

Goodyear Trophy in May 1952, finishing fifth, and in the National Air Races in July 1952, finishing third in the Grosvenor Trophy.

CHAPTER XIII

Pilot Fatigue

In 1961 there developed among BEA pilots, particularly those flying the Comet, a feeling that their flying duties had become too arduous. There had been several accidents and 'incidents' which could have been caused by the pilot or crew being over-tired. The Ministry therefore immediately set up machinery to investigate this serious problem – the 'Flying Personnel Research Sub-Committee into investigating Pilot Fatigue in BEA'. The Chairman appointed was a retired RAF medico, Group Captain Ruffel-Smith, who had considerable experience of dealing with pilots in the RAF – notably those engaged in Bomber Command operations during the War. The other members of the sub-committee were a BEA doctor, a doctor (female) specializing in psychology, and me. My role was to represent the pilots' angle and advise on the general working and operations of an airline. We had a number of meetings at which we discussed how best to handle the inquiry. The first problem was to identify the factors likely to cause or contribute to fatigue. They covered a wide field, viz.:

1. Actual flying time.
2. Duty time – i.e. flight time, plus the time necessary for pre-flight briefing, meteorological briefing, preparation of flight plan and its clearance, post-flight debriefing.
3. Journey time to and from home.
4. Time of day.

5. Intensity of workload – i.e. number of radio calls, changes of airways etc.
6. Number of aerodromes at which approaches, landings and take-offs had to be made.

We allocated points to each of these. We then obtained a timetable of all the routes flown, and arranged that we would individually fly each route, virtually as a supernumerary member of the crew. In the Viscount cockpit there was a spare seat, known as the 'jump' seat, which we were able to occupy, and from which all activities of the crew could be monitored. We equipped ourselves with a comptometer on which we kept count of the number of radio calls made and frequencies changed. We strapped a pad on our knees for notes. If the schedule involved a night-stop, we stayed at the same hotel as the crew, and generally participated in any off-duty activities. I recall one schedule which I thought might be rather arduous. It involved taking off from London at 10.20 for Geneva – flight time 1 hour 22 minutes – along busy airways. Stop for 59 minutes, than take off for Athens, flight time 2 hours 40 minutes. Then take-off for Nicosia, flight time 1 hour 33 minutes, followed by 52 minutes on the ground; then take-off for Beirut, flight-time 46 minutes, arriving Beirut 19.31. This gave a journey time of 9 hours 7 minutes, plus flight preparation and clearance time, about 2 hours; plus travel time from home to the airport – and from the destination airport to the hotel, say $1^{1}/_{2}$ hours – a total day of nearly 13 hours. One would have thought that the crew would have been looking a bit tired towards the end of their duty, but, on the contrary, they seemed to get more lively! The reason, I soon found, was that they loved Beirut! In those days it was known as 'the Paris of the Middle East'. We stayed at a very good hotel called the Commodore. In the basement was a fine restaurant and nightclub. The crew could not get out of their uniforms quickly enough, into civvies – and down to the bar. Of course it was part of my job to go with them!

Among the flights I made was London-Rome-Athens-Istanbul, where we night-stopped. Pilots were occasionally allowed to bring their wives with them, and on this flight the Captain and First Officer had availed themselves of this concession. After dinner I was invited to go with them to have a look at the nightlife of Istanbul. As often happens we 'men' were walking together and the two wives were some few yards behind. We had turned down a street at random and found we were in the 'Red Light' district, with glamorous females beckoning from open windows. Suddenly there was a commotion behind us, and we turned round to see the two wives being threatened by several of the 'ladies of the town'. Then a couple of policemen appeared. The ladies of the town demanded to see the licences of the wives, these being a legal requirement in Istanbul. Despite our attempts to intervene, the wives were taken to the police station and summoned to appear in court the next day for appearing to be plying for their trade without a licence! We all went along to the police station; we spoke no Turkish and no-one there spoke English. After much gesticulation, eventually forms were produced, which the wives were asked to sign. They were applications for a prostitute's licence!

With some trepidation the wives signed them. They were then asked for the fee, given licences, and with much smirking we were all allowed to go. I heard later that the wives had the licences framed, and that they now occupy an honoured place among the mural decorations of their respective homes!

As a result of our investigation we came to the conclusion that the average workload on pilots was too high, and therefore likely to cause fatigue. Taking the calendar month as a basis, we totalled the points which the average pilot earned. We then reduced this total by an arbitrary amount, and recommended that when a pilot had reached this total of points in any month, his flying duties should cease. Our recommendations were accepted and put into action. We heard that implementation of them had resulted in some pilots using up their points in three weeks or so, and thus being off flying for the remainder of the month. To the best of my knowledge the points system is still in use.

I enjoyed this experience, not only for the further insight it gave me into the workings of an airline, but also for the opportunity it gave me to renew acquaintance with pilots I had met before, and to get to know others.

The experience of serving on this sub-committee investigating pilot fatigue is one that I shall always remember.

CHAPTER XIV

United Nations Technical Assistance Mission

In May 1953 my secondment to the Flying Unit – and my job as Commandant of the airport – came to an end, and I was posted back to Ministry (MCA) HQ. We had by this time bought a nice old thatched collage at Sheering, Essex, not far from Bishops Stortford. I became a commuter, catching the 8.15 from Bishops Stortford each morning, and getting home about 6 p.m. My job was Deputy Director of Training and Licensing. Standards were still in a relatively undeveloped state, and much of the work entailed meetings with operators and with visits to airlines, charter operators, flying clubs etc. There was also continuing liaison with the International Civil Aviation Organisation of the United Nations (ICAO) on the standardization internationally of licensing and safety requirements. In line with this, the UN had set up a Technical Assistance Bureau – UNTAB – to assist emergent and underdeveloped countries to put their civil aviation on a safe and uniform basis. In June 1956 I was asked if I would like to work with the UN on a secondment basis. I said yes, and was assigned to take over an international mission of experts in various fields – telecommunications, airworthiness, operating standards – from various member states: Holland, France, Canada, Greece. It was to be known as 'The Middle East Air Safety Project'. Our first base was Damascus, Syria.

Sue and I packed our bags and reported to Heathrow, there to board a BOAC aircraft for Damascus. We were met by a small delegation from the Civil Aviation Department and taken to a hotel, where most of the team were already assembled. There followed meetings between the Mission, as it was called, the Aviation Department, and Syrian Airways, at which we planned how to set about our task. The first step was obviously to inspect and check their existing organization and their standards of flying, telecommunications, airworthiness etc. I was responsible for the flying and licensing side, and in this capacity flew with them on their various routes.

I recall one of my early flights; I had control in the co-pilot's seat when the aircraft became rather tail-heavy. I handed over control to the Captain and walked aft. I soon discovered the reason – almost all the passengers had moved to a clear area near the door – and were sitting on the floor round a primus stove, making coffee!

Life in Damascus was interesting. The Souk in the old part of the city fascinated dear Sue – and me. The 'Street called Straight' was exactly as it was when St Paul was let down in a basket from an upper window. We had a modern flat with a stream cascading through the garden, which was fed by an aqueduct constructed nearly 2,000 years ago by the Romans. At weekends we frequently went over to Beirut in Lebanon. It was a glamorous place in those days – known often as 'the Paris of the East'. There was a British club where we used to stay called the St George's, situated right on the Mediterranean, with a natural swimming pool formed by the rocks.

But we were to move again, dramatically. Trouble had been growing between the Arabs and the Jews, in what was to be known as the 'Suez Crisis'. Feelings against Britain began to run high. One morning I found written in the dust covering my car: 'DETH IS FOR YOU'. I rang Montreal for advice for my

Mission. They said they would attach a Colonel from the UN Peacekeeping Force in Lebanon, to assess the situation hour by hour. He came over and issued us with 'walkie-talkies'. We were to listen out at fixed times. If we got the message 'Evacuate' we were to get into our cars and quietly go to Beirut. I had a problem here – I had just purchased a car from a chap in the Embassy who was returning to England. It still had its UK number-plates. Very fortuitously one of my Syrian staff remembered seeing a UN number plate in the basement of our office. We fitted it, but of course had no supporting documentation. The order 'Evacuate' came, and off we set, Sue and Alastair and myself, with a few essentials, wondering whether we would ever see Damascus – and all our belongings – again.

Approaching the frontier I was on tenterhooks lest they should ask for the car identification papers. However, after a hard look at the distinctive blue UN plates – and a look at our UN *Laissez passer* – we were waved through.

Entry to Lebanon – a very relaxed country in those days – was simple, and we duly reached Beirut, where we had booked rooms as usual at the St George's Club. But anti-British feeling soon spread. One day Sue and I were walking along the beach, when some locals started throwing stones at us. Our HQ ordered a further evacuation of the mission – to Athens. We of course had our cars to dispose of. Fortunately the UN Peacekeeping Force HQ had a flat roof used as a car park. We were allowed to leave our cars there. Would we ever see them again?

Despite the circumstances of our travel we were all very intrigued at the prospect of a stay in Athens as guests of UN. We were accommodated in the Grande Bretagne Hotel, one of the best hotels, and the subsistence allowance which we collected from the local UN HQ – in US dollars – was generous. One day I recall sitting with Sue sipping an oozo, with the backcloth of the

Dining at the Seven Brothers, Athens.

Parthenon not a mile away. We got into conversation with a local who spoke English. 'What a wonderful place the Parthenon is,' I said.

'I suppose so,' he replied. 'I've never visited it!'

Our stay at Athens did not last long. One day we received orders reassigning the Mission – to Israel!

Our HQ were at Tel Aviv. We were met by a small delegation from the Civil Aviation Department and accommodated in the best hotel. During the next few days we sought permanent houses and found a very pleasant first-storey flat with a balcony commanding a view of the town and its surroundings. We soon realized that the Israelis were already fairly well organized. Their national airline – El Al – covered routes to the USA, where of course many Israelis had emigrated over the years, and where El Al was highly respected. They were the first airline to adopt the Bristol Britannia, closely followed by BOAC. For some time El Al held the record for the fastest crossing of the Atlantic by a passenger airline. They had several British pilots, and a very good Operations Manager named J.E.D. Williams – Jed – a former navigator of RAF Transport Command. In addition to El Al, there was an internal airline which mainly served Eilat, a very popular place at the end of the Red Sea. We went there frequently and enjoyed the myriads of wonderful fish which were one of its main attractions. My other favourite place was Caesarea. It was much as it had been for hundreds of years. We used to have the beach largely to ourselves. At one place was a natural pool in the rocks known as Cleopatra's pool, and history relates that she and Antony disported themselves in its waters. We found several Roman coins and other relics of its earlier days.

The city of Jerusalem was divided between Israel and Jordan. The old city, surrounded by its walls, and containing most of the places of interest, was on the Jordanian side. There was a stretch of about 200 yards of No Man's Land between the two halves, entered from the Israeli side by the Mandelbaum Gate. With our United Nations *Laissez Passez*s we were allowed through at any time, and frequently crossed the frontier to spend a weekend in the Old City, visiting the places we had so often heard about in our religion and reading of the Bible. It had changed little in its 2,000-odd years since the Roman occupation and earlier. The narrow stone-flagged streets bore witness to the traffic of chariots; today, untroubled by modern traffic, the donkey, camel and bicycle hold sway. Dear Sue was fascinated by the many open-air stalls selling brassware, jewellery, embroidered work, oriental rugs and food, the smell of which was never far away from one's nostrils. We were particularly attracted by the many old and colourful rugs from Persia, Afghanistan, and other romantic-sounding places. Much haggling went on over the price. The merchant never expected you to pay the first price asked, and obviously enjoyed the bargaining. Sometimes, if we liked a particular rug, we would 'leave it', and saunter back on our next visit. The merchant would greet us cheerfully, and if the rug was still there, we would usually agree a price. Our houses have been graced by these rugs and carpets over the years, a reminder of the interesting times we had.

One of our disappointments in our time in Syria and Israel was that Sheila was not with us. She had won a scholarship to the Herts and Essex High School in 1955 and was doing well, showing particular interest in music. We felt that it would be a great pity to interrupt this important period in her life. It so happened that the school had made provision for a number of boarders, so Sheila became a boarder. She was always able to go to her 'Granny's' at Towcester or West Kirby for short breaks. For the longer holidays she came out to us. Here, my contacts with civil aviation operators proved useful. Skyways, whom I knew well, operated a trooping contract to Cyprus, and readily agreed to bring Sheila out. I would go over to Cyprus to meet her. During our time the ENOSIS trouble started in Cyprus. One day I went over to meet Sheila arriving at Nicosia airport. While waiting, I opened a door and stepped out onto the tarmac. The next minute I felt a bayonet poking into my ribs, and I was ordered peremptorily to get back. Then I heard that the aircraft had been diverted to a Royal Air Force airfield. I got through to them and was informed that they had 'Miss Donaldson' there and would send her over by road to Nicosia. She would arrive in about an hour's time. I waited at the entrance anxiously, and within the hour a small convoy of three jeeps arrived. Sheila was in the second jeep – with an armed convoy in front and behind! Something to tell the girls about when she returned to school!

We enjoyed our time in Israel, but alas our assignment came to an end, and we had to say farewell to the many nice people we had met and worked with. There were farewell parties, and gifts which we still treasure.

During our time in Israel we had bought a car under the duty-free concessions to which we were entitled, a Ford Zephyr. We thought, Why not drive back to England? Overland all the way presented some difficulties, so we decided to go by sea – from Haifa to Naples. On arrival at Naples, our first visit was to Pompeii, then on to Rome and Venice and many other interesting places, finishing at Ostend, where we boarded a ferry to Southend – and so home to our cottage at Sheering. The date was August 1959.

After a few weeks leave I reported back to my department, to be informed that I was posted to Northern Division as Divisional Operations Officer and Deputy Controller. The job entailed visiting all the aerodromes in the Divisional territory, which included Northern Ireland and the Isle of Man. Our HQ was in

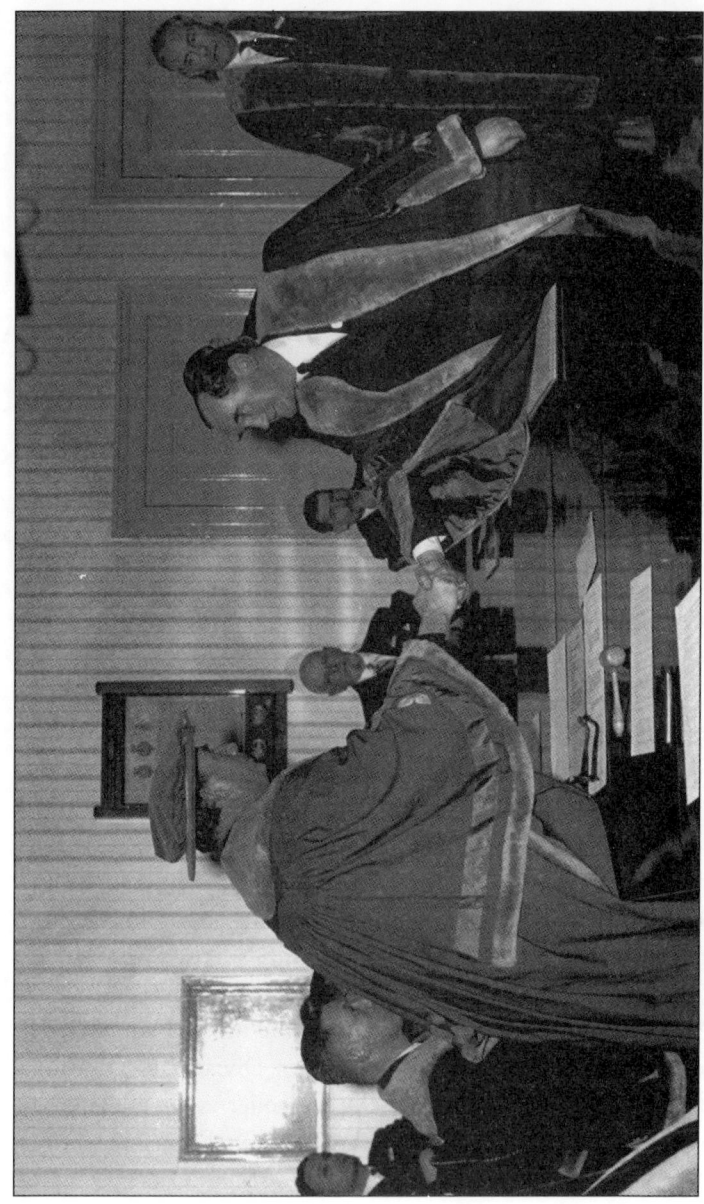

Receiving the Freedom of the City of London, 26 May 1960.

Liverpool. As my mother and sister lived in West Kirby, they said, 'Come and stay with us, and commute.' I accepted their kind offer, but I was not happy with the life, involving as it did absence from my family, so I asked for a posting back to London where we could resume our life together at 'The Old Cottage', Sheering. My request was granted but I was posted, not to HQ in London, but to London Divisional HQ at Heston, near London Airport. This involved a journey from east to west across the prevailing north-south traffic into London. I soon found it was not practicable, and was faced with the options of Bed-and-Breakfasting from Monday to Friday in Hounslow or some such place, or moving house. With reluctance, we decided on the latter course. We had loved our 'Old Cottage', and would be leaving some good friends.

On the home front, dear Sheila, now sixteen, had been pursuing her musical bent, and had won a scholarship to the Royal College of Music. The move should not unduly affect her movements. Anyway, she had ideas of joining with several friends and renting a flat in London. Alastair was now a boarder at Clare School in Somerset. In retrospect I do not think I handled his education correctly. I had been obsessed with the determination that he should go to a good public school. This stemmed from my own experience. I had been at a local secondary school. My close friend in my late teens had been at Fettes College. I found myself rather on the fringe of a set, most of whom had been to well-known public schools – Sedbergh, Shrewsbury, Ampleforth, Stoneyhurst – and I suffered a feeling of inferiority in their company. It is interesting to recall that at that time dances were mostly private, organized by parents. A dance outside that category was called a 'Subscription Dance'. Tails and white tie were the order of the day. A dinner jacket was *de rigeur*. My feelings were perhaps further developed when I joined the RAF as an officer. The majority of fellows in these days were from public schools. Thus I had made up my mind that Alastair should never suffer similar feelings if I could help it. I made application to Fettes and Uppingham, only to find that there were long waiting lists. Eventually I got a place for him at a fairly new school, Clare School, in Somerset. But he was not happy.

Back to the move from Sheering. I was faced with the problem of where to look for a house, reasonably accessible to my office at Heston, but not in suburbia as represented by the surrounding districts. I thought about Maidenhead, and set off after work one evening with several agents' brochures. I was returning along the Bath Road when I noticed a sign-post to the left – Burnham Beeches. I had heard of Burnham Beeches as an interesting beauty spot, so, having nothing to do, I turned left to explore. It was a lovely area of woodland, interlaced with roads arched by trees, and with the occasional stretch of water populated by swans, geese and other birds. I eventually came to a signpost pointing to Farnham Common. Entering the village at a T-junction, I saw a signpost indicating Beaconsfield to the left, Slough to the right. I had not gone more than a hundred yards or so, when I noticed a house on the right with a 'For Sale' board up. It was an interesting-looking old house, of Queen Anne vintage I hazarded, with a number of outbuildings. It was 9 p.m., but I thought, Well, they can only say 'Sorry, see the agent'. I entered the gate and knocked on the door. A lady opened it a few inches, and asked my business.

'I saw your "For Sale" board,' I said, 'and wondered if I could make a few enquiries. I appreciate it is rather late, and will understand if it's not convenient.'

She was attired in a rather scanty dressing-gown.

'No,' she said, 'wait till I slip something on, and you may come in.'

It was an old farmhouse which had once farmed many acres in the area and been surrounded by open country. It retained its

Mead Farm, Farnham Common.

Dear Sue.

character: a range of single-storey outbuildings; a two-storey stable block and a wall surrounding the farmyard. The house comprised a kitchen/breakfast room, scullery, dining room, a large entrance-hall and a drawing room – all oak-beamed. The drawing room commanded a view of a lawn and garden of about quarter of an acre. The breakfast room had a door leading out to a flagged patio, with a summer house. 'How much are you asking for it?' I enquired.

'Eight thousand pounds,' she replied.

I liked it. I brought dear Sue over; she liked it. We became the owners of 'Mead Farm' as it was called, and were to spend twenty-three happy years there. The year was 1963.

CHAPTER XV

Farnham Common – Aviation Consultant

We settled in after the usual problems associated with moving house, and I started commuting the eleven miles or so to my office at Heston. The journey was greatly assisted by the recent construction of the M4 motorway which accounted for some eight of the eleven miles. I had first become acquainted with Heston in 1938 when it had been hailed as the new London Airport, and had been the home of my Reserve Flying School contractor, Joe Birkett. But development around it had not been controlled, and when the time came after the War, it was decided that there was insufficient open space for expansion – and so it was abandoned. When it was constructed, it had been provided with a very fine terminal building. The Department decided that this would make an excellent HQ for Southern Division. I therefore found myself in possession of an office overlooking the deserted airfield which I had used to visit twenty-five years before.

One day there was great activity on the airfield. Bulldozers and mechanical shovels appeared. They had found that there was a valuable deposit of clay under the ground. For the next few months this clay was excavated and carried away, leaving a large pit. The pit filled with water and my view was improved – seagulls, and occasional geese and swans became visitors. But alas one day large trucks appeared, full of refuse and rubbish! They had decided that the old airfield was to be used for housing development. So the pit had to be filled in. What better filling than refuse and rubbish, the disposal of which was always a problem?

After lunch one day, I took a walk over to the pit to see if there was anything of interest; it is amazing the things that people will sometimes throw away. I noticed some very nice pieces of York stone crazy paving – then a whole load of it! I took my car over each day and collected it, taking as much as possible home each weekend. I relaid it in my garden, around a pool I had constructed and it made a very attractive feature.

In 1969 I was posted back to HQ, commuting by train from Slough to Liverpool Street. Shortly after moving into our new home, we received our first caller. He introduced himself as Len Brockwell and said, 'I hear you were in the RAF?'

'Yes,' I replied.

'So was I,' he said. 'We have recently formed a branch of the Royal Air Forces Association, and hope you will join.'

I joined and it became one of our main interests. enabling us to meet and make friends with people who 'spoke our language'.

'Did you know so-and-so? He was out in India,' was a frequent line of conversation. We were fortunate as a branch in that the Farnham Common-Beaconsfield area was the dormitory of pilots flying from London, Heathrow Airport. In those days 99 per cent of commercial pilots were ex-RAF. I remember one of these, a senior captain in British European Airways, saying. 'You won't remember, but you checked me out as an instructor.' He brought his old flying log book along. In the remarks space of his assessment I had written, 'Aerobatics require more polish.'

Apart from the camaraderie we enjoyed, our function was welfare. We organized concerts and events to raise money for our funds. During Battle of Britain Week there was great activity, doing house-to-house collections. I would set out at about 5.30

and get back to a late supper with dear Sue, at about 9 o'clock – sometimes later, if the itinerary included a pub!

One concert which we organized stands out in my memory. It was staged at Cliveden, a beautiful mansion, the home of Viscount Astor. Here many notable people, including Sir Winston Churchill and Lord Beaverbrook, had been frequent visitors. It later achieved great notoriety as the venue of the Christine Keeler-Profumo affair. Sheila, then on the threshold of her musical career and a student at the Royal College of Music, helped to organize the show, and sang in it herself. Dear Sue and the other wives laid on wonderful refreshments. Over the year we would raise £4,000-£5,000 for our charity and for welfare. I was privileged to become first President of the branch – and still am!

Several of our new friends had tennis courts. Geoffrey and Dora Ashwell invited us to their tennis party every Sunday. Then on Thursdays Minnie Halliday had a tennis afternoon – for girls only. Sue adored her tennis – I can see her now skipping across the courtyard in her white shorts, racquet swinging, getting into the car to join the 'girls'.

Then there was the Farnham Common Sports Club, catering for tennis, badminton, cricket, hockey – and rugby. I took an interest, particularly in the rugby team which called itself 'The Drifters'. My playing days were of course over, but I followed the team keenly from the touchline. I recall one afternoon when the referee did not turn up. 'George, would you referee?' said the captain. 'I'll have a go,' I replied. My knowledge of the finer points and their interpretation was rusty, so I thought the best approach was to let the game flow. In the evening, over the traditional pint, I recall one of the team sidling over and saying, 'It must have been fun playing rugby in Tom Brown's schooldays!' They presented me with the referee's whistle I had used, on its ribbon. I still have it. I blow it sometimes.

My retirement was now approaching, and Sue and I used to discuss what we were going to do. A cottage in Spain or Devon or Cornwall? We decided against going overseas, but thought we would explore the West Country. We hired a caravanette and drove by easy stages to Lands End, calling at many of the well-known holiday resorts on the South Coast. From Lands End we took the route along the Bristol Channel Coast, and one day found ourselves in Bath.

We had always been interested in antiques, and indeed thought it might be a pleasant pursuit in our retirement. We came across an antique shop which we entered. In the course of conversation with the owner, we mentioned one of the purposes of our tour. He said, 'I wonder if you would be interested in a cottage which my sister is selling – it's at Ston Easton, about six miles away?' We visited it and liked it, though it was somewhat remote. We were not ready to move yet, but thought that it might be a good investment. It was cheap – £2,800 – so we bought it. We spent odd weekends at our country cottage, but could not quite see ourselves spending the rest of our lives there. We let it to a young couple who didn't pay the rent, so we sold it, and were back to square one.

Our interest in antiques was growing; we realized that the two-storey barn adjacent to the main road would make an ideal showroom, so we started to collect antiques. We found it interesting going to auction sales, particularly where the sale was the contents of an old house. One seemed to catch the atmosphere of the bygone age represented by the various artefacts. We also found it very enjoyable to go out for the day visiting other antique shops. If we liked a particular piece, we would ask the price, then announce that we were 'trade' – we had cards printed – and usually get a reduction. I made a sign 'ANTIQUES' which swung on a wrought-iron bracket over the pavement outside.

Soon customers started to call. As in our own case, some of

them were 'trade'. Some of them came from overseas. There were American military bases in the area; 'Gee, isn't that just cute!' became a frequently heard expression. There were callers from Europe, particularly from Germany and Holland. One day a German couple called and bought a few items. They came again about a month later. It was evening. (Like the Windmill Theatre in London, we never closed.) After concluding our business, we said, 'Would you care to come in and have a little refreshment?'

'Thank you, we would very much like to,' they said.

After settling down with a glass of whisky, Klaus – for this was his name – said, 'You must be about my age – what did you do in ze War?'

'I was in the RAF,' I said.

'Ah so,' he replied, 'I was in ze Luftwaffe!'

We shook hands, and there began a friendship with him and his wife Barbara which lasts to this day. I took them up to our RAF Association Club where they were well received. The next time they came over, they brought a handsome set of beer tankards – Steins – which they presented to the club. Klaus became a regular caller and friend. Then there was Yon, from Amsterdam. He still calls on Alastair regularly and buys items which Alastair has collected for him.

In the course of our business I had purchased an old Ford van which had been advertised locally for £50! I was still travelling to London daily, on the 8.33 from Slough, which was about three miles from our home. I had a friend living a few houses away with whom I used to travel. Quite frequently dear Sue would drive us to the station in the van. It was always a source of amusement to us to see the faces – and double-takes of fellow-travellers as the doors of the van were opened by glamorous Sue, to release two gentlemen attired in dark suits, wearing bowler hats and carrying well-furled umbrellas and briefcases!

131

Klaus and Barbara Roser, clients who became great friends.

Retirement age in the Civil Service was sixty, however I was invited to take a two-year extension. Apparently the ravages of time had not bitten too deeply. I accepted and quite enjoyed my final years. I eventually hung up my bowler hat, and, with the usual ceremonies, retired. The year was 1970.

With retirement we were able to give more time to our antiques. Sue became quite interested in upholstery. She enlisted on a course which was conducted by the local authority night school. She became very good at it, and, in addition to re-upholstering chairs and settees which we had acquired, received many calls for her expertise.

When we moved to Farnham Common, Alastair was fifteen and a boarder at Clare School. He was not happy there and we at last accepted the fact. We therefore took him away and sent him to the local Slough School as a day-boy. He never quite caught up with his rebellious years academically, but he soon revealed a bent and talent for practical work and the use of his hands. He bought an old Morris car which became his joy and pride, and there was nothing he could not do in the way of repair and renovation. But it was in carpentry and in the restoration of furniture that he showed his true talent. He virtually took over my workshop, built up a set of fine tools and did many repairs to our antiques, while building up his own business. One of the pieces of furniture in our house which is often admired is a Welsh dresser; it was one of Alastair's earliest works, the traditional top being made and matched to an old pine chest-of-drawers! Furthermore, he realized that study – of techniques, of period furniture, of the various woods etc. – was important, and built up a useful technical library.

Sheila meanwhile was pursuing her musical career. She had graduated well from the Royal College of Music in singing, and later with a teaching diploma. She joined an operatic company, and toured the country, playing in such well-known shows as *Hello Dolly* and *The Merry Widow*. But she decided that a future 'living out of a suitcase' was not for her, and that perhaps a career solely in music might not yield an adequate living. So she took a course in speech therapy, in which she qualified, becoming a speech therapist with a local authority. But she became disillusioned, finding that she was employed in other duties in addition to those for which she was engaged. Her next job, which she enjoyed was as a receptionist to a practice of dentists in Harley Street. She met many well-known people – MPs, duchesses and so on – and became a valued 'partner' in the practice. But then she found an interest in china restoration. She took a course in it and adopted it as her profession, whilst still continuing with her music, singing in churches and at concerts – and teaching.

Now, back to Farnham Common. I had retired, and my dear Sue and I were contemplating an easy life. But it was not to be. One day I received a letter from the Department. There was to be a public inquiry on proposed development of the Leeds/Bradford Airport. The inspector needed a technical assessor. Would I like the job? I thought it would be interesting and Sue agreed. So I took it. This was the time for airports to be developed. Newer, larger aircraft demanded longer runways, and new runways. Noise was becoming a problem. People living around airports objected; public enquiries became the norm. After the Leeds/Bradford Public Inquiry it seemed that I had become accepted as 'The Technical Assessor', and between 1969 and 1976, I took part in inquiries at the airports of Leeds/Bradford, Manchester, Glasgow, Edinburgh, Gatwick, Luton, Stafford, Hounslow, Bristol, Fair Oaks, and in respect of proposed heliports at Shadwell Basin on the Thames and at Hemel Hempstead. Apart from the fact that I was kept away from home sometimes, I enjoyed this work very much. I was fortunate in working with inspectors with whom I got on very well. This was important as at the end of a day we spent most of the evening discussing evidence – usually over dinner.

At the Gatwick Inquiry, on the second day, I was dining with the inspector, whose name was Gates. As we talked, he quietly slumped forward. I thought he had dozed off; but he did not respond. I called a waiter and told him to summon a doctor. I felt his pulse but could feel no heartbeat. We laid him on the floor and I tried mouth-to-mouth resuscitation until the doctor and ambulance came. They took over and applied oxygen – but he was dead.

It was now 11 p.m., and with the Court due to resume in eleven hours, I realized I had to do something – but what? Gatwick was under the British Airports Authority, the Chief Executive of which was Sir Peter Masefield. I had known Peter over the years, during which he had progressed through the Civil Service and British European Airways to his present post. Indeed, he had been present at the inquiry. I phoned his home; fortunately it was he who answered. After expressing his sympathy and shock, he said, 'Leave it to me, George.' He must have worked very hard, for I was duly informed that the inquiry had been adjourned, but would continue after one day's break. The next day a new inspector was assigned. I met him and briefed him as fully as I could – and the inquiry resumed.

In Scotland the system was slightly different. There was no special section of the Government with a staff of inspectors. The person in charge of a public inquiry was selected from barristers in the top rank. They were known not as inspectors, but as the 'Reporters'. My reporter for the Glasgow Airport Inquiry was a tall, bespectacled Edinburgh barrister named Bennett. The inquiry had been convened because of objections to a proposed runway extension by people living in the area whose 'quality of life' would be adversely affected. Our final recommendation was that the extension should be allowed.

Shortly after the Glasgow Inquiry came the inquiry at Edinburgh. It was for similar reasons – a proposed extension to the main runway would affect the residents in one of the higher-class suburbs. The reporter was again a well-known Edinburgh barrister, Gimson by name. In the course of our getting acquainted prior to the opening of the inquiry, it emerged that he had served in the Army against the Japanese in the Far East. He had been taken prisoner-of-war after the fall of Singapore and had been forced to work on the notorious railway between Siam and India which the Japanese were constructing in the furtherance of their grandiose plan to control the world. I told him of my involvement; of how our last assignment, after the atomic bomb had ended the War, had been to drop supplies of essential food to them, and to fly them back to civilization. He recalled the immense joy and relief they had all experienced when our Dakotas appeared and parachutes fluttered down divulging cooked chicken, cereals etc. – and, not least, crates of beer! He said of his flight back from Don Muang Airport, Bangkok, 'You know, George, I thought your face was familiar. It could well have been that I flew back in your aircraft!' The rapport which this war experience evoked probably prompted him to invite me to stay with him. He lived alone in a rather lovely flat in an old Georgian mansion. I accepted his kind invitation, which made my stay more enjoyable, and greatly helped our discussions.

On the opening day of the inquiry I suddenly spotted the tall, bespectacled figure of John Bennett, who had been my reporter at the Glasgow Airport Inquiry. After warm greetings, I said, 'What are you doing here, John?' 'I'm Council for the Objectors,' he replied. 'I live in the affected area!' I couldn't help a chuckle, recalling his approval of the extension at Glasgow! A day or so later John invited me to dinner. I said it might perhaps be unethical in my 'neutral' position to dine with counsel representing one of the parties, so I asked Gimson, who of course knew Bennett well as a fellow-barrister. He could see no reason why I should not go. The objections to the extension were not supported by the Court's findings; John Bennett still remained a good friend – as did Gimson.

During the Leeds/Bradford Inquiry I met a Dutchman in the bar of my hotel. His name was Hans. In the course of conversation he told me that he was very keen on rugby, and that it was becoming popular in Holland. I told him that I belonged to a club called the 'Drifters' in Farnham Common, and said, 'Why don't you bring your club over and play us?' The following weekend when I was home, the phone rang; 'This is Hans speaking from

Holland. My club would like to come over to play the Drifters –
at Easter.' That was about two weeks away. 'Good' I replied. 'I'll
enquire if they can arrange it.' I rang the secretary and he said, 'Yes
– I think we've got a free day on the Saturday.' I rang Hans and
the visit was arranged.

They arrived in a coach – twenty of them at our house on the
Good Friday. They were a lively bunch, obviously looking forward
to a good weekend. Sue accommodated two of them, and the rest
were soon invited home by other members. The weekend was a
great success and concluded with a cordial invitation to the
Drifters to come to Holland the next year. They did, and the
match became a regular event. The Captain of the Drifters, Jamie
Davidson, eventually married the sister of one of the Dutch team.
Strange how the lives of others can be influenced by a random
event in one's own!

The East Midlands Airport came into being on the site of a
wartime aerodrome, Castle Donnington. After the War, Castle
Donnington also became a motor racing circuit, and as such was
to achieve world fame. A conflict of interests arose. The main
runway of the airport was aligned so that the majority of take-offs
were towards the racetrack. The owners of the racetrack wanted to
develop the spectator facilities. The owners of the Airport objected
on the grounds of safety, and wanted to be free to extend the
runway in the future. An inquiry was convened to consider the
various interests. High among them was, of course, safety. The
airport maintained that to allow spectator development in the
racetrack area immediately beyond the end of the runway would
lead to severe casualties in the event of an aircraft crashing
immediately after take-off.

One day I received a call from a great friend and aviation
colleague, Ron Gillman. He was a member of our RAFA branch
who had had a gallant career as a pilot during the War, and

subsequently became a senior captain in British European Airways. He said, 'George, I've been approached by the developers of Castle Donnington Racetrack to see if I can help them. They are involved in a public inquiry on developments at East Midlands Airport. Will you help me?' I had not yet been invited to be the Technical Assessor, and indeed that was not an assumption one could ever make. I was rather intrigued with the prospect and said I would. I received an official invitation to act for the Castle Donnington Racetrack Authority.

It seemed to me that our evidence should be directed at convincing the inspector that the risk – of an aircraft crash-landing in the racetrack area and causing serious casualties – was no greater than at many other airports in the world. To this end I carried out detailed research and obtained particulars of some thirty airports, including such important places as Hong Kong, at which development was more likely to be a hazard than that of the racetrack was to East Midlands Airport.

The assessor appointed was an old colleague of mine in the Ministry – George Mackintosh. We couldn't help the odd smile at each other. The evidence which Ron and I presented satisfied the inspector that the developments planned for the Castle Donington Racetrack were an acceptable risk: the Airport's objection was not allowed.

Airport inquiries petered out; my 'job' as a 'technical assessor' also petered out. It had been an interesting experience.

CHAPTER 14

'D and H'

In the course of her household duties at Mead Farm my dear Sue – a wonderful housekeeper among her many other qualities – employed a 'daily help', Mrs Hudd. One day, at about the time of my retirement, I said, 'I must get someone to help me to do a bit of house decoration.' Mrs Hudd, who was scrubbing the kitchen floor, stopped and said, 'Excuse me, Mr Donaldson, but my husband is quite good at that, and I'm sure would help you in his spare time.' 'Ask him to come up, Mrs Hudd,' said I.

Bill, for that was his name, came up. I liked him and took him on. In the course of conversation, I asked him what his normal work was. 'I'm a lagger,' he said, 'or, to give the job its posh description, I'm an insulation engineer. The work is putting various types of insulation on boilers and pipes to conserve the heat.' He went on to tell me that he worked for a contractor, that there was plenty of work about, and that he only wished he could be in business on his own. I thought about what he had told me. Why shouldn't I start him up on his own? When an opportunity arose, I mentioned that I might be interested in putting up a little capital to enable him to 'start up on his own'. He was very pleased. We held a number of meetings and I went into the implications of starting a business. They did not seem too difficult, so we decided to go ahead. We formed a limited company with a nominal capital of £1,000. He became the working manager, and a director. I became the Managing Director. We decided to call the firm 'D and H Insulation Ltd.'

Bill completed the work on which he had been currently engaged, and handed in his notice to his employer. I set up an office in one of our outbuildings, framed our Certificate of Incorporation, had our notepaper and billheads printed etc. and we started to look for work. As he had said, Bill had a few contacts; we advertised, and I contacted firms in the Slough area with which I had some acquaintance. The work started to come in.

I soon realized that I had gravely underestimated the various ramifications the work involved. I had envisaged a few hours a day in my office, while Bill got on with the practical side. He soon told me he needed help, so we advertised for 'Insulation Engineers'. We engaged one, and so immediately had to face insurance and other requirements. Bill did not get on with him, so he left, and thereafter we employed a number of 'laggers'. Alastair even had a go, but soon concluded it was not for him. So Bill said, 'I think we'll just undertake the work I can do myself, Mr D.' (Throughout our time together, he never called me anything but 'Mr D'.) This did not prove practicable, though, and he often used to say, 'I wonder if you would come over and give me a hand, Mr D?' Thus more and more I found myself being drawn into the practical side.

There are broadly two methods of insulating tanks and pipework. One, by fixing pre-formed slabs of insulating material on large surfaces such as tanks, and pre-formed 'sectional' around pipework. Two, by applying insulating plaster. Sometimes a combination of both was called for. Sometimes a sheet-metal covering was added. The first method needed no skill, and I was therefore able to join in such work without any trouble. Thus in time it came about that I found myself:

1. Salesman,
2. Estimator,
3 Clerk and Typist,
4. Accountant,
5. Practical Lagger
   – all in addition to the overall running of the business!

Quite a large proportion of jobs entailed the removal of old lagging. This, even if not pure asbestos, frequently contained a large proportion of that substance. Asbestos was just becoming recognized as dangerous – highly detrimental to health. Strict regulations were made controlling the conditions under which work was done, wherever asbestos was present. Work had to be screened off from adjacent areas; workers had to wear masks and protective clothing; special changing rooms, with shower-baths had to be provided. There were additional complications which I had not anticipated, and did not relish. But I had 'set my hand to the plough...'

A few jobs stand out in my mind. There was one near Pontypridd in the heart of the Welsh Valleys. It involved insulating six large cylindrical tanks about eighteen feet high by ten feet in diameter, to be used to hold certain substances from animal hides. The final work was to be encased in stainless steel sheeting. The latter required considerable research to obtain the sheeting and to determine how it should be fitted I found that standard sheets could be obtained six feet by three feet. The obvious method of attachment was by rivets, so we had to buy a drill and riveting tool. Scaffolding was necessary. Then there was the problem of getting each sheet in place. I eventually devised a system employing a small hand-hoist which could be moved around the perimeter of the tank. The main cable divided into two strands fitted with clips which attached to the sheet. I would attach the clips and then operate the hoist under Bill's guidance from above. Obviously it made it easier to pre-drill the eight rivet holes in each sheet. I was doing this one day in the warehouse of the supplier, when an employee sidled over to me and said, 'Are you Union, mate?' My immediate reaction was to say, 'No' which might have had embarrassing results, as it was a strong Union area. Then I suddenly remembered that, as a Civil Servant, I had become a member of the Institute of Professional Civil Servants – the IPCS – and that at one of our meetings some of our more militant members had found that we were eligible to become affiliated to the Trades Union Congress, the TUC. A motion to join had been put to the meeting, and passed! I therefore said confidently, 'Yes.'

'What's your union'?'

'IPCS.'

'Never 'eard of it,' he said.

'Well, why not check up?' I replied.

Somewhat undecidedly he turned on his heel and left. I heard no more.

One of the larger firms which frequently gave us work was called 'Planned Maintenance'. They in turn were under contract to the Public Works Department. Windsor Castle was in their area, and one day we were given a job there on boilers and pipework under the Castle. It was fascinating walking along old passages and into cellars which went back to Norman times. One of the exits came up into a tower which formed an archway over the road up to the Castle. We ascended from the depths one day via this exit and were sitting on some steps having our lunchtime break, when a gleaming Rolls Royce approached and stopped. We jumped to our feet and stood respectfully. In the back was the Queen. An aide rolled down the window and the Queen said, 'What are you doing here?'

'We're insulating boilers and pipes in the cellar, Ma'am,' said I.

'Oh, that's good,' she replied. 'I've been a bit worried about the size of our heating bills lately.' She smiled, and the car resumed its journey. We heard afterwards that she took a great interest in the domestic affairs of the Castle, and frequently paid unheralded visits to the staff.

One day I got a call from Mrs Hudd – 'Bill doesn't feel too good this morning, Mr D, so he won't be coming in to work. It's his bronchitis.' Bill's bronchitis didn't improve. and his doctor advised that he went into hospital for further examination. I visited him daily. His condition got no better. About five weeks later he died. The cause of death was – asbestosis.

I was very saddened by Bill's death and concerned at the cause, wondering if our business was in any way to blame. But it emerged that Bill had been exposed to the dangers of asbestos for many years and the disease had been developing for a long time.

I placed 'D and H' into voluntary liquidation and settled up all its affairs. I was very sad at Bill's demise and its effect upon his family, but I was not sorry that this chapter of my life had closed.

Thus in my 'retirement' I had been working with 'D and H', participating in public inquiries as they came along, and helping Sue with her antiques. Yet we still seemed to have time to enjoy our lives together. We extended our acquaintance with the Mediterranean, spending holidays in Crete, Sicily, Rhodes, Greece, North Africa, Spain and Gibraltar. The last-mentioned brings back special memories. One of my old friends and colleagues, Brian Oakley, had retired to Spain. He and his wife acquired a flat in Algericas, overlooking the small stretch of water to Gibraltar. I recall that on his retirement he had asked, not for the traditional silver salver or coffee-pot, but for a pair of binoculars. He envisaged spending his time sitting out on his balcony watching the shipping in the harbour and the aircraft flying in and out of Gibraltar airport. Sue and I spent one or two pleasant holidays with him. Then, sadly his wife died. Brian's health began to fail, and it became too much for him to entertain us. The long-standing trouble caused by the question of the sovereignty of Gibraltar had eased, and the frontier with Spain had been opened. We therefore thought we would combine a holiday in Gibraltar with being able to visit Brian 'across the bay'. It proved to be a happy arrangement. But we did not like Brian living alone and did our best to persuade him to return to England, where we would be able to look after him – he had no children, alas, and no close relatives. But he would not be persuaded. I recall his saying, 'I cannot say I am happy here, Don, but I am content.'

Gibraltar's strategic role as the gateway to our Empire in the East had declined, but it was still a military garrison with a Royal Air Force element. I therefore thought there might be a Welfare Section, which could keep an eye on Brian. I found that there was a unit of the Soldiers, Sailors and Air Forces Association, so I paid them a visit and told them about Brian, who had of course been a Wing Commander. They were very helpful and said that they would arrange for their Welfare officer to visit him at regular intervals. The Welfare Officer was a WAAF, and thereafter I used to receive regular reports from her. One day I received a phone call from her telling me that Brian had passed away. She seemed very upset and I like to feel that, over the period she had been visiting him, there had grown up between them a warmth of feeling deeper than that normally called for in her welfare capacity. Sue and I were saddened at the loss of our old friend.

CHAPTER XVII

Kith and Kin

I was born in England, of Scottish parents. My father came from
Edinburgh – Leith, actually; my mother from Greenock. I had
two sisters, Agnes eight years older and Jane six years older than I.
We were brought up to be a God-fearing family; to go to Church
every Sunday, and take an interest in church activities. Father and
Mother sang in the church choir; Father's interest in music
extended beyond the Church. He was for many years a member of
the Liverpool Philharmonic Society Choir, and trained a choir
made up of girls in the area, which won a number of
competitions. By profession he was a mechanical engineer and
draughtsman, but he had a natural bent for most things practical.
I remember him making an 'express cart' for me, and a sword for
my role in 'Robin Hood' at school. He made wireless sets in the
early days. He was a very good carpenter – I still have a bookcase
and other items that he made. He cultivated a successful
allotment, and Mother seldom had to buy vegetables. Mother was
a good housekeeper and needlewoman – she and Father were very
devoted. Mother had a brother, Tom, and two sisters, Margaret –
'Aunt Maggie' – and Elizabeth, 'Aunt Lizzie'. We used to go up to
Greenock most years for our summer holidays. In those days there
was a shipping line, the Burns Line, which plied between
Liverpool and Greenock. We used to travel by it, sleeping in a
cabin overnight. I recall that in August 1914 we were en route,
entering the River Clyde, when we were hastily got out of bed and

145

told to assemble in the main lounge. The Captain came in and told us that England was at war with Germany, and that German submarines were known to be in the Clyde estuary. We were given lifeboat drill and issued with lifejackets, but we were not torpedoed.

My sisters, Nessie and Winnie, as she was then called (I never knew why, as she was christened Jane, and reverted to this name in later life), were quite different in character. Being the eldest, Nessie seemed to accept a responsibility for me and my upbringing. In later life she accepted it as her responsibility to look after Father and Mother, and then, after Father died, to look after Mother. I think it was this sense of duty which contributed to her remaining a spinster. Nevertheless Nessie lived a very full and rewarding life outside the home. She became Secretary to the Director of the Walker Art Gallery in Liverpool, a post which she held for thirty-eight years. This must have generated an interest in art beyond the duties of secretary. She learned about artists, and it became an accepted part of her work that she lectured schoolchildren in an appreciation of art. She took up painting herself, becoming a reasonable artist – a respected member of the Deeside Art Group – leaving many paintings, some of which have decorated our homes and those of her friends over the years. In her earlier days she was a keen Sunday School worker. She was selected by our church to go on a residential course at West Hill Training College, which qualified her to change radically the old system in which children of all ages were lumped together, and introduced the 'Graded System', in which they were divided into 'Beginners', 'Juniors', 'Intermediate' and 'Senior'. She became keen on Folk Dancing, attending seminars and weekend 'meets' in various parts of the country. She even got me to participate on one or two occasions, pushing or pulling me as the sequence required. She sang in the local church choir. She was also a very active member of the

Womens' International League for Peace and Freedom; and during the war she served in the Women's Land Army! I helped her to move to a flat nearer to the shops. She retained her interests and her circle of good friends until she died in 1987 at the age of eighty-eight. I missed her very much. Back in West Kirby I still meet people who remember her fondly.

Jane and I were more of a kind – on the rather rebellious side. Without appearing to be studious, Jane won a scholarship to Liverpool University and obtained an Honours Arts degree with Honours in English Literature. She then went up to Oxford and gained a teaching diploma. She taught at a school in Wokingham, but not for long. She also studied for a time at Leipzig University. But she met Leonard Woodward and after a colourful courtship married him. Len was a brilliant scholar. He took an Honours degree in Chemistry at Lincoln College, Oxford, in 1922, and a career in Imperial Chemical Industries was at his feet. But he preferred the academic life and returned to Oxford, initially as a tutor but finishing up as Vice-Principal of Jesus College, with many other distinctions. He and Jane set up their home in Oxford. They were blessed with a daughter, Jean, who became my 'favourite niece'! In our courting days Sue and I were always welcome, and their home was the frequent venue for Christmas celebrations.

I recall an occasion during the War when I was night-flying and got caught in a fog. Radio navigational facilities were not very good, but I knew I was somewhere near Oxford. Eventually I said 'Here goes' and started to descend, hoping to find clearance. At 600 feet, just as I was about to climb back into the murk and try somewhere else, I found a clearance. Then we spotted a beacon flashing out its code identification. 'KI' – that must be Kidlington,' I said to my navigator. He confirmed, and we circled the aerodrome and landed thankfully. Kidlington was just a few miles out of Oxford, so I rang Jane to enquire if she could put us up. She said 'Yes, of course.' We borrowed a jeep and found our way to her house. After the usual greeting and introductions I said, 'Where's Len?' Jane replied, 'Oh, he's sleeping in College these nights. He's engaged in some rather vital work to do with the War. We're working on it with the Americans. We believe the Germans are working on something similar. Whoever gets there first will probably win the War!' About twelve months later, when I was in Rangoon waiting to take part in the invasion of Malaya, the atomic bomb was dropped on Hiroshima and the War was over. I realized then what Len had been doing. Len died in 1976, and Jane shortly afterwards. Len never forgave himself when he realized fully the awful purpose to which his research had been put.

Jean followed in her mother's footsteps to some extent, getting a scholarship to St Andrews University and obtaining a degree in Zoology. She married a Cambridge graduate in Chemistry, Jack Hayden. Unlike Len, he entered the business world, starting a successful partnership as a consultant. Their home is in Maidenhead; they took over the mantle of their parents and we spent many pleasant family Christmases with them. They were blessed with two fine children Nicola and Christopher. Nicola has married and has been blessed with a daughter, so dear Sue and I now have a great-niece!

Dear Sue was born and bred in Towcester, an ancient Roman town on the historic Watling Street. Her father was a dairy farmer, and Sue loved to tell how she frequently had to deliver the milk before breakfast! Her mother – always called Cissie – was a lovely person, very straight, but full of fun. Though untrained, she had a gift for music. She only had to hear a tune once to be able to sit down at the piano and give a very close rendering of it. Sue had a brother, Bill, a year or so older, a motor salesman. He was a nice fellow who married a girl, Evelyn by name. Evelyn was the dominant partner. She ran a successful hairdressing business, perhaps more successful than Bill's motor-salesmanship! Sue never liked her. From my small acquaintance, neither did I. Bill wanted children, she didn't – so they were childless. She died and for a few years before Bill died too, we were able to repair the family relationship.

Sue had three cousins. Joyce is the daughter of her mother's sister, Aunt Elsie. She is happily married to Trevor Sweet. They have twin girls and a third daughter. Pat and Peter are the children of Aunt Betty, another of Sue's mother's sisters. Pat's first marriage ended in divorce. She then married an American and moved to Los Angeles where she now lives. She has no children. Peter worked in his father's bookmakers business for a time, but tended to move from one job to another. He married and they had a son, but his wife died; the son became somewhat estranged, possibly due to the lack of the traditional influence of a stable home. Peter visits his sister Pat in Los Angeles frequently. He likes it there and has now virtually severed any links he had with England to live with her. He has a caravan and apparently spends much time in it getting to know this vast and interesting country.

Dear Sheila and Alastair have settled, apparently contentedly, in the homes they have bought in Kington. Sheila has her ever-faithful friend Michael living in an adjoining house, and her boyfriend George visits her from London periodically. She is kept busy in her profession of china restoration, continues with her music, singing professionally in the Parish Church in Hereford and at concerts, and having pupils in singing. She has become a Samaritan, for which I admire and am proud of her. Alastair pursues his craft of antique restoration, always having time to help his sister or friends in the many practical jobs which arise.

Sheila and George, and Alastair have kept in daily touch with us, bless 'em, and have visited us periodically, which has given us much pleasure.

In 1984 my dear Sue, who had always been such a fit and healthy girl, complained of pains in her body and legs. She had to give up her tennis and badminton, and for a time her symptoms puzzled her doctor. But eventually they were diagnosed as Polimyalgia Rheumatica. It was a great blow to us. Eventually it 'burned itself out' and Sue returned to some semblance of her former self. But, alas, she was stricken with cancer, and had mastectomies first on the left and then on the right breast. She bore these with great gallantry, and I am ever grateful to God that this foul disease did not return. But, in 1988 she began to forget and lose her ability to do the things which she loved doing. Her condition was diagnosed as Alzheimer's Disease. We found that we required two qualified nurses, but even with these the strain became too great, and with reluctance I was persuaded that dear Sue would have to go into a nursing home.

We found a nursing home and they agreed to take dear Sue. Then Alastair found a bungalow within 200 yards of the Home. I inspected it, and though it was not my ideal, I realized that it had much to offer, and I acquired it. I spent most of each day with dear Sue, taking her for walks down to the town for a bit of shopping, and to the bungalow. I built a conservatory out into the garden, and constructed a pool with a fountain. Here we would sit. I like to feel that it gave her some pleasure. I would give her tea, then at about 7 o'clock we would have a little drink together, as we had done for many years.

On 18 April 1997 my darling Sue passed away peacefully. We had been together for sixty-two happy and eventful years. Life will never be the same again.

My dear mother and father brought us up to go to church and to participate in its life. I have, alas, not always followed its teachings, but for the last few years I have become increasingly aware of God, and believe in His power for good in our lives. Tennyson once said, 'More things are wrought by prayer than this world dreams of.' I have found that this is true.

CHAPTER XVIII

Epilogue

In 1989 we sold our beloved house, Mead Farm, Farnham Common. Sheila and Alastair had left the fold; it was bigger than we needed; and we felt that we could put the capital tied up in it to better use. We considered buying a smaller property in Farnham Common or its environs, but decided that if we were to take full advantage of the sale of Mead Farm we should seek a house where prices were relatively lower. Nessie's flat in West Kirby was still on the market. We had let it, furnished, while trying to achieve a sale. Why not give the tenants notice, and live there pro tem? By a stroke of good fortune the tenants found the house they had been seeking and moved out of the flat. He was a distinguished research chemist – Dr Owen Standen by name. He and his charming wife, Joyce, were to become two of our greatest friends.

As Nessie's flat was fully furnished, and we had no intention of making it our permanent home, we decided to store our furniture – collected over the years and very dear to us as it was. Here we were fortunate. Alastair's house in Kington possessed stabling and several commodious outbuildings. Why not store our furniture with him; he and Sheila could probably 'give some of it a home' anyway? So this was done – and we duly moved to Nessie's flat in West Kirby.

As a place to live West Kirby has much going for it – and is indeed much sought after. I had of course known it well from my early days. But it was with no pleasant feeling of nostalgia that I returned. Dear Sue had never taken to it. Now with her health deteriorating, she did not care. As the fell disease with which she had been stricken took its course, the need for nursing care increased – until we needed it throughout the twenty-four hours.

Knowing that we were not happy, dear Sheila and Alastair proposed that we move to the Kington area. We were very grateful for this concern, and decided to explore the possibilities. There was never any question of living with either of them; this would have been too much to ask. The ideal would be to find a cottage or flat with accommodation for live-in carer, and the availability of other temporary nursing care. The alternative would be to find a good nursing home, and convenient living accommodation nearby for me. Initially we had to put dear Sue in a nursing home about six miles away, while I stayed with Alastair and explored the availability of accommodation for us both.

In July 1996 there was a room in the nursing home in Kington, so we moved there and installed dear Sue. Alastair found a bungalow for me not more than 100 yards from the Home, and he and Sheila fitted it up nicely. I spent most of the day with dear Sue at the Home. When the weather permitted, I would take her out in the wheelchair. We would venture down the town or come back to the bungalow, where I would give her tea, and we would have a little drink about 7 o'clock – as we had done for years. On 18 April 1997 my darling Sue passed away, peacefully. We had been together for sixty-two eventful and happy years. Life will never be the same.

Many years ago I remember King George the Sixth, in his Christmas Day broadcast, offering a prayer which appealed to many people; it went:

Oh Lord support us all the day long of this troublous life,
Till the shadows lengthen, the evening falls,
The busy world is hushed, the fever of life is over
    And our work is done.
Then, in Thy great mercy, grant us
A safe lodging, a holy rest,
    And peace at the last.

177 Wing – a Short History

No. 177 (Airborne Forces) Wing, Royal Air Force was formed on 1 October 1943, at Rawalpindi, Punjab, India. Its function was to provide aircraft and aircrews to carry troops of the 50th (Indian) Parachute Brigade into battle in the SE Asia theatre of war.

The Wing initially comprised three squadrons of Dakota aircraft, Nos. 62, 117 and 194. The nominal strength of aircraft in each squadron was twenty, plus one or two reserves. The establishment of aircrews was about two per aircraft. The make-up of a crew was pilot, navigator, and two wireless operator/air-gunners, who were to add 'Jumpmaster' to their qualifications. Squadrons were fortunate in having a strong Commonwealth element. Picking a squadron at random. I found that, out of a total of 161 there were 20 from the Royal Australian Air Force, 22 from the Royal Canadian Air Force, 5 from the Royal New Zealand Air Force, and 1 from the South African Air Force.

The 50th (Indian) Parachute Brigade comprised No. 152 Indian Battalion, No. 153 Gurkha Battalion, and No. 154 Gurkha Battalion, plus Brigade Signals Section, and No. 411 (Royal Bombay) Parachute Section, Indian Engineers.

The Brigade had completed its basic training, and had been waiting some time for operational parachutes and aircraft with which to train for battle. Priority was low, due to the war in Europe and the Middle East. Eventually the C-in-C India,

General Wavell, sent a personal signal to the Prime Minister, Winston Churchill:

> My Parachute Brigade has been starved of help and equipment from the U.K., and much effort will be required if it is to play the very valuable part it might.

The signal had now yielded results.

I met the Brigade Commander, Brigadier Tim Hope-Thompson. at the earliest opportunity. Battalion commanders got together with squadron commanders and a good liaison began to develop between our two units.

Apart from No. 62 Squadron, which had been based at Chaklala for several months, assisting No. 3 Parachute Training School, the squadrons had little if any experience in paratroop dropping. They therefore underwent a short period of training, during which the WOp/AGs added 'Jumpmaster' to their qualifications, and pilots learnt the drill for dropping a stick of paratroops into the dropping zone (DZ). All of the WOp/AGs and most of the pilots and navigators also did the ground paratroop training course, followed by a 'live' jump.

Most of the troops had done their basic training in old Vickers Valencia aircraft, dropping through a hole in the floor. These were replaced by Hudsons, which were not very suitable and only accommodated eight troops. The Dakota was, therefore, a great improvement, accommodating twenty fully-equipped paratroops, who ejected by jumping out of its wide doors.

Thereafter operational exercises were discussed and planned, starting with a few aircraft and working up to exercises involving the whole Brigade, and the sixty-odd aircraft of the Wing.

The aim was to reach a standard whereby the Brigade could be dropped on a map-referenced point up to 400 miles distant, at a specified zero-hour. It required a plan on the 'count-down'

Fiftieth Anniversary celebrations of the formation of 50 (Indian) Parachute Brigade and No. 177 (Airborne) Wing, RAF, Agra, India, 1992.

principle, the most critical section being the time to be allowed for navigating and map-reading to the DZ, particularly at night, and perhaps by an indirect route.

With our Airborne Force approaching readiness, we attended conferences at various Army HQs at which operations were discussed. At one of these, I recall, Lord Mountbatten took the chair. The use of the force in two operations in the Indaw area of North Burma looked promising. They were aimed at assisting General 'Vinegar Joe' Stillwell and his Chinese forces. Alas, both had to be abandoned, owing, I think, to the inability to obtain General Chiang Kai-Chek's unconditional agreement. The most important one would have been BULLDOZER, a combined Land-Sea-Air operation to recapture Akyab. But this also had to be abandoned, for various reasons which are not for me to relate.

It was now the beginning of 1944. The Japanese forces had advanced to the eastern frontier of India. If they defeated our Army here, the road to Delhi would be open. All squadrons of the Wing were ordered to the Indian/Burma Front. The 50th (Indian) Parachute Brigade (less one battalion) frustrated, also moved east to join the conflict, not in their proper role as paratroops, but as light infantry. They were to play a gallant part in the forthcoming battle. We were to meet again.

After their long retreat through Burma, our forces were reorganizing, and had started an advance in the Arakan. But they were meeting with formidable opposition from the Japanese, who had succeeded in cutting communications from the north. It had been planned from the start that the 81st West African Division, in the Kaladan River area to the east, would be entirely on air supply. Now the remaining three Divisions, Nos. 5, 7, and 26, were virtually cut off; and General Slim had called for all-out air supply.

The requirements of an army in action, and their packing ready for air-supply, were the responsibility of the Army. The loading and despatch of the packages from the aircraft were the responsibility of the RAF, though volunteer 'kickers out' from the Army were welcomed, and did a fine job. The Dakota's maximum useful load was 7,500lb (three and three-quarter short tons) adjustable in relation to the fuel required for the flight. Loads with parachutes were ideally dropped from 600'–700'. Sometimes the nature of the terrain prevented aircraft going so low. 'Free' dropping – of items which were not unduly affected by impact – was usually from about 100' although quite frequently parachutes failed to open. When this happened and the load was ammunition, the result could seem quite unfriendly! There was one case when an aircraft landed back with petrol dripping from the fuselage. In his debriefing the Captain said, 'We found several of the cans were leaking, so we thought we'd better bring them back!' He must have been an Irishman.

After dropping their load, aircraft would land at the nearest landing-strip and pick up casualties. A typical report reads:

Picked up 21 cases of Smallpox, 2 of Cholera, 3 died in plane.

The Arakan area was some 200 miles from Base supply airfields at Comilla and Agartala. A sortie of this nature could take four-and-a-half to five hours. Quite often crews would do two sorties in twenty-four hours.

During February 1944, covering this phase, our aircraft dropped 3,933 tons, mostly by night to avoid enemy action. An extract from one of the squadrons' operational records for 10 February reads:

From 1545 hours on 10th, to 0845 hrs on 11th, 19 aircraft flew 36 sorties. Our aim had been 47, but the Air Supply section had been unable to keep up.

After their initial success in this sector, achieved by bold and aggressive tactics, the Japanese attack was repulsed. The troops of XV Corps fought back with great spirit, and routed the Japanese Army in this sector. General Slim wrote of it:

> The Arakan battle, judged by forces engaged, was not of great magnitude. But it was, nevertheless, one of the historic successes of Arms. It was the turning-point of the Burma Campaign.

We were proud to have helped.

Wing HQ was now established at Agartala. On arrival we were augmented by No. 31 Squadron, which had already distinguished itself in supply-dropping operations for Wingate's first expedition into Burma. Nos. 62, 117 and 194 Squadrons were disposed at airfields in the vicinity, accommodated in hastily-constructed 'bashas'. Conditions were primitive, but everyone remained cheerful and keen. Later we were augmented by a detachment of No. 216 Squadron from the Middle East, bringing our total strength to around 1,000 Dakota aircraft.

With the information of a unified command in SE Asia, under Lord Mountbatten, a measure of integration between US and British forces was decreed. At our level it was decided that 177 Wing should integrate with four American C47 (as they called the Dakota) squadrons, to form 3rd Troop Carrier Command. We were given offices in the same block as General Slim, commanding 14th Army, and Air Marshal Sir John Baldwin, AOC 3rd Tactical Air Force, under whom our new command came. It was a happy and workmanlike arrangement, enabling developments to be discussed, and action initiated immediately. At squadron level, the integration made little if any impact. Brigadier-General 'Bill' Old was a keen and gallant pilot, who preferred to be flying to sitting at a desk. I don't remember ever getting a 'minute' from him – even a 'loose' one. I see from my logbook that I flew 150 hours during February/March, so there must have been times when both offices were vacant. We were fortunate in having a very good Senior Air Staff Officer in 3rd TAF, Air Vice-Marshal Gerald (later Sir Gerald) Gibbs, who did all the work.

Here I feel it would be helpful if I attempted a broad review of the military situation, so that the work of our squadrons may be put into perspective. As outlined earlier, IV Corps were now holding a front to the south, in the Arakan area, extending roughly from Kaladan, westwards to the Bay of Bengal. From Kaladan, the front ran generally NE for some 700 miles, to the Fort Hertz area, where General 'Vinegar Joe' Stillwell held precarious sway with his Chinese forces. The intervening territory, poorly mapped, consisted largely of thick jungle, rugged mountain ranges, rivers, and swamps; many areas were there for the taking by either side if they could negotiate the hazards, the mountain ranges and main rivers in the north and south, thus forming a natural barrier between Burma and India. To the west of this barrier, 14th Army was deployed – the main force at Imphal, in a valley of that name some 20 miles long by 10 miles wide, surrounded by mountains up to 8,000' in height. Virtually the only supply route into the valley was from the railhead at Dimapur in the north, via Kohima – a distance of 130 miles, over a mountain range, across rivers and through tortuous valleys.

General Slim realized that the primary objective of the Japanese must be the destruction of his 14th Army. Faced with no simple options he decided:

1. to withdraw his force deployed to the north and south into the valley, and
2. to air-lift 5th Indian Division from the Arakan into the valley.

Thus he hoped to achieve maximum concentration, and possibly numerical superiority of his forces. Furthermore, he would be fighting on ground of his choosing, against an enemy whose communications would be extended.

But it was not to be quite as he had hoped. The Japanese probed west more rapidly and more fiercely than had been anticipated. To the south they cut off 17 Division's withdrawal route to the valley, making it entirely dependent on air-supply. More seriously, to the north they cut the valley's only supply route from the railhead at Dimapur, at a point 30 miles north of Imphal. The date was 30 March 1944. On 5 April, 17 Division gallantly broke through the Japanese road block to the south, to join the besieged forces in the valley. On 17 April, Kohima, with a garrison of 3,000, was surrounded.

Among the forces opposing the Japanese around Imphal were our old friends, the 152nd and 153rd Battalions of the 50th (Indian) Parachute Brigade. At a place called Sangshak they were attacked by vastly superior numbers of the enemy. They held their ground despite a grave shortage of supplies including water. Alas, many of our supply-drops fell outside the DZ. However, the historian, K.C. Praval, in his *Indian Paratroopers*, quotes this report by the commander of the 152nd Battalion:

> One aircraft however came very low and made a number of runs, so that we were able to collect the entire aircraft load. The pilot was magnificent. Each time he made his run, he was so low that Japs opened intense fire on him, while we waved and shouted encouragement. He was so low that we could see him waving at us, and also the Despatchers in the doorway. All subsequent supply-drops were of the same pattern. On making enquiries afterwards, I learnt that the pilot and crew had taken part in the Brigade's air-training at Chaklala. On hearing that the 50th Brigade was cut off, and having to rely entirely on air-supply, they determined that whatever happened, and regardless of risk, the Brigade should get their entire load.

General Slim wrote of this engagement: 'The ten days delay and the heavy casualties this small force, and the R.A.F. who supported them, had inflicted on the enemy, were of inestimable value at this critical stage of the Battle.'

Despite this tribute, I have always regretted that aircraft of the Wing could not have done more to alleviate the desperate plight of our old colleagues.

Apart from XV Corps' recent success in the Arakan, there had been little to cheer about in the struggle against the enemy. However, a man named Wingate had caught the imagination by a bold venture in which he marched a force into Japanese-held Burma, disrupted enemy communications, skirmished with the enemy, and, battered but undaunted, marched out again. Now he wanted to repeat the operation. But this time he wanted his 3rd Indian Division to be flown in. The operation, to be code-named THURSDAY, was approved. The 177 Wing element of 3rd Troop Carrier Command was detailed to provide forty aircraft to assist. The full and absorbing story of this unique operation has been well portrayed in other accounts.

When I interrupted my review of the general situation to describe the Chindit operation, it was 7 April. Imphal was besieged; Kohima was besieged. Thus our tasks were:

1. The air-supply of the Kohima garrison.
2. The fly-in of reinforcements and supplies to Imphal.
3. The air-supply of the Chindits.
4. The air-supply of our forces on the Arakan front.
5. The movement of personnel and equipment of RAF tactical squadrons to other airfields.
6. The evacuation of casualties.
7. The collection and distribution of mail.

The siege of Kohima was given priority by General Slim, because to have allowed it to fall without the stoutest resistance would have made our supply-base at Dimapur more vulnerable. Thus we engaged in intensive supply-dropping to the 3,000 strong garrison there. The town was encroached upon by the enemy, until the garrison was confined in a space roughly 500 yards square. All aircraft were having to run the gauntlet of ground fire, yet it is surprising that no aircraft were brought down or seriously damaged, though many were holed or received slight damage. I recall that we received a visit from an armament officer at HQ Delhi. He said he would like to accompany us on a sortie. I told him he could come with me as a despatcher. We had just completed our drop when one of the regular despatchers popped his head into the cabin and said 'Squadron leader Paul's been hit Sir.' 'Where?' I replied. 'In his bum, Sir.' Fortunately it was not serious, but he returned to Delhi somewhat chastened.

The Siege of Imphal was probably the turning point of the war in SE Asia. In order to exist and fight, four divisions of mixed British, Indian and other nationalities' troops had to be supplied by air with all essentials. The operation was code-named STAMINA.

The Army's initial bid was for 400 tons per day to be flown in. Sir John Baldwin, in consultation with his staff, and Troop Carrier Command, accepted the bid.

My recent research of old records has brought home to me the expertise which the Army had developed in gauging and meeting the problems of supplying an army. Here, for example, were the daily and weekly scales of rations for British and Indian troops. Some of the differences intrigued me. The bulk of Tommy's daily ration was, 'Bread, 14 ozs', as against the Sepoy's, 'Rice, 18 ozs' Both were allowed $^3/_4$ oz of salt. Tommy got $^5/_8$ oz tea and $3^1/_2$ oz sugar against the Sepoy's $^1/_3$ oz tea and $2^1/_2$ oz sugar. Tommy was allowed fifty cigarettes per week, as against forty for a Sepoy. Both were allowed two boxes of matches, unless you were a non-smoker, when you were allowed one. Tommy was rationed to thirty-five sheets of toilet paper per week; a Sepoy – none!

By this time the monsoon was developing, and the 8,000' high mountains surrounding the valley were frequently shrouded in thunderous cumulo-nimbus clouds and heavy rain. On many occasions it was impossible to fly into the valley. Kumbirgram was an airfield just to the west of the mountains – about twenty minutes flying time from Imphal. So it was decided that supplies should be stockpiled there awaiting clear weather, when all available aircraft would concentrate on a rapid shuttle service. This arrangement was code-named HUSTLE.

I recall being at Kumbirgram one day, and noticing a Dakota taking rather a lot of runway before getting airborne. A few minutes later it landed back. I went over in my jeep, to be met by the pilot complaining that the aircraft would not climb. Just then a flight sergeant drove up, took one look through the fuselage door, and said, 'No f— wonder – you've got a double load!' The aircraft had been carrying sacks of *ata* (grain) a full load of which, weighing 7,000lb odd, would be three layers in depth. This aircraft had six layers! The maximum all-up weight of the Dakota at that time was 31,000lb. We reckoned that this one had taken off with about 37,000lb – and landed again safely. A great tribute to the Dakota.

The Siege of Imphal officially lasted from 18 April to 30 June 1944. During this period, in addition to other duties, aircraft of the Wing transported 4,399 tons of stores and equipment into the valley – and brought many casualties out.

On 22 June, the vital road to Dimapur had been reopened, and Kohima had been relieved. By the end of June 14th Army was breaking out of Imphal on the offensive. After their heroic operations into enemy territory, most of the Chindits had been withdrawn or flown out for well-earned rest and recuperation. Those units remaining merged into 14th Army's new offensive.

Third Troop Carrier Command was officially disbanded on 4 June, and 177 Wing reverted to its earlier status. The role of the Wing remained the same – but we were now supporting an army on the offensive.

Although the diversity of our tasks might appear to have been reduced, in fact our work became more exacting and intensive. In their move eastwards and southwards, the Army were having to penetrate this mountainous, jungle-covered terrain, which had from time immemorial formed an almost impenetrable barrier. Virtually all supplies had to be dropped. Dropping-zones were continually being moved and were often difficult to find. The approaches to many were hazardous. Distances from supply-bases were increasing. And, overall, the monsoon was at its height. Nevertheless, in the ensuing months, squadrons increased their flying hours, one squadron taken at random totalling 2,770 hours in June. This was a great tribute, not only to aircrews, but to all our ground personnel who seldom allowed our aircraft availability to fall below 90 per cent.

By September/October, 14th Army had cleared all pockets of enemy resistance around Imphal, and had advanced eastwards and southwards to a line roughly from Tamu in the north to Kalemyo in the south, some 150 miles from Imphal. The Japanese had resisted fanatically all the way, as was evidenced by the massed bodies of dead found in villages overrun. By now our squadrons were being relieved for rest and recuperation, which usually took the form of training at their old stamping ground in the Punjab. I was posted to 229 Group as G/C Operations towards the end of August, handing over command to Group Captain G.N. Warrington. On 30 September 1944, No. 177 Wing was disbanded.

During its comparatively short existence, the Wing had trained for 3-4 months to carry a parachute Brigade into battle – but without the satisfaction of doing so. Its role had then changed to the support of 14th Army and squadrons of the 3rd Tactical Air Force, in virtually the whole range of their needs, from Bofors guns to butter, from mules, horses and oxen to toilet paper, from petrol and ammunition to drinking water, from motorcycles to jeeps, from Spitfire tanks to aero engines, from battle-trained troops to the evacuation of the sick and wounded. Of the last-mentioned, General Slim wrote: 'Air evacuation did more in the 14th Army to save lives, than any other agency.'

Our aircraft provided an air-mail service which kept the troops in touch wherever they might be. The operations record of one of the squadrons reads: 12th May. F/O J.G. Simpson crashed 'Clydeside' with 6,000 gallons petrol; and mail. Escaped, but went back to aircraft – now blazing, and rescued mail-bags.'

In their 8 months of operations on the Burma Front, aircraft of the Wing transported 33,610 tons of animals, stores and equipment, 31,217 troops, and evacuated 23,898 casualties.

Fourteen aircraft were lost due to enemy action; nine due to weather, terrain and other causes; many were damaged by enemy fire.

Sixty-one aircrew were killed, and four wounded.

The relatively small scale of our losses was in large part due to the fighter escorts of Hurricanes and Spitfire always available from 3rd Tactical Air Force. Sometimes four, sometimes six aircraft would closely escort a sortie. We owed them a great deal. I should like to take this opportunity of thanking them.

The ubiquity of the Dakota in coping with its many tasks and the conditions in which they were carried out was only matched by its ruggedness and simplicity. These qualities, together with the skill and dedication of our maintenance crews, were paramount in the carrying out of our job.

Gallantry awards included twenty-three DFCs, one bar to the

First Day Cover – Fiftieth Anniversary of the end of the Second World War.

168

DFC, one BEM, and one DFM. Additionally our American allies honoured us with six American DFCs.

War seems to be a necessary evil, leading inevitably to the destruction of human life. Few people can but abhor this aspect. It is perhaps some source of comfort that the operations of 177 Wing were in no small part directed to the sustaining of life, and the succour of the sick and wounded.

Extract from *194 Squadron Royal Air Force –*
The Friendly Firm by Flight Lieutenant Douglas Williams,
Merlin Books Ltd, Devon, 1987, pp. 22-3.

Before we left Basal there was a Wing Inspection and an amusing incident happened to one of our ground staff, Peter Donaldson, who with his chum Andy Anderson, was on detachment at Chaklala, to make an engine change and billeted in the Victoria Barracks at Rawalpindi. In the barracks one afternoon, lying flat out on his charpoy (bed), Peter buried in a thriller, suddenly froze. A loud voice was bellowing in his ear: 'Corporal, stand up.' Peter told us as follows:

'I looked up to see the whole barrack room at attention. My bed was surrounded by a group of officers and the Warrant Officer who had shouted. The officer in the centre had gold braid on his hat. Here go my two stripes, I thought. "What's your name, Corporal?" "Donaldson, Sir." There was a sudden quiet and then from the gold braid, with a smile, "Well keep awake, Corporal."'

The inspection over, Peter enquired from his mates who the hell that had been.

'Why,' the answer came, 'Group Captain George Donaldson.' Knowing George, I am sure he may well have thought, There but for the grace of God go I. George became a very good friend of 194 Squadron and is today a proud honorary member of our Squadron Association. George was on the officer selection board at

Peshawar in October 1943 when I was up for a commission and I well remember him saying to me with a twinkle in his eyes, "Ah! I see you are a Welshman, can you say that long Welsh name?" Whereupon I replied nervously, "I have been a long time away from Wales, but I will have a go," and much to my astonishment reeled it off pat. George showed obvious delight and said, "I thought it ended with OGOGOCH," or words to that effect. I often wondered if I got my commission on the basis of that eloquent Welsh pronouncement and the fact that I played rugger, as my bank account was not particularly healthy at that time, a state which would have immediately disqualified me in peacetime.'

Other mentions of GKFD occur on pp. 19-20, 24, 30, 35, 50, 55, 65, 67 & 76.

Reconciliation

In March 1993 I received a phone call from the Ministry of Defence. They said that they had been approached, through the Japanese Embassy, by the Japanese Broadcasting Company, who were producing a feature on the War in South-East Asia. They had never been able to understand why, after carrying all before them – Shanghai, Singapore, Malaya – they were halted and finally defeated in Burma. They were told that it was largely due to the fact that our forces had achieved superiority in the air. This was vital to the transport of troops and equipment, and their supply from the air. It was suggested that they get in touch with the commander of 177 Wing, which, in co-operation with the United States Air Force, had operated the hundred or so Dakota aircraft involved. So it was arranged that they come up to see me. The delegation included cameramen and lighting- and sound-engineers. Oddly the person in charge was a young – and very attractive – Korean lady. I was asked to give my account of the campaign, and to respond to a number of questions. I was later sent a copy of a video they had made. I did not think it was very good, but it is interesting that the Japanese have admitted their failings and sought the reasons.

At Easter I received a card from the four-year-old son of the producer who had been with his mother when the recording was made. I thought that a nice gesture. I think he is in the cockpit of the aeroplane.